AF587662

BIOCHEMISTRY RESEARCH TRENDS

LIGANDS, POLYMERS AND AMINO ACIDS

BIOCHEMISTRY RESEARCH TRENDS

Additional books in this series can be found on Nova's website under the Series tab.

Additional E-books in this series can be found on Nova's website under the E-book tab.

BIOCHEMISTRY RESEARCH TRENDS

LIGANDS, POLYMERS AND AMINO ACIDS

TSISANA SHARTAVA
EDITOR

Nova Science Publishers, Inc.
New York

For permission to use material from this book please contact us:
Telephone 631-231-7269; Fax 631-231-8175
Web Site: http://www.novapublishers.com

LIBRARY OF CONGRESS CATALOGING-IN-PUBLICATION DATA

Ligands, polymers, and amino acids / Editor, Tsisana Shartava (Tbilisi, Georgia).
p. ; cm.
Includes bibliographical references and index.
ISBN 978-1-61122-793-2 (hardcover)
1. Biochemistry. I. Shartava, Tsisana.
[DNLM: 1. Polymers. 2. Amino Acids. 3. Ligands. QD 381]
QP514.2.L545 2011
612'.015--dc22

2010047635

Published by Nova Science Publishers, Inc. † New York

CONTENTS

PREFACE

This book presents and discusses a variety of research done in the field of biochemistry, with a particular focus on ligands, polymers and amino acids. Topics discussed include G-Quadruplex Ligands and the evaluation of a synthetic biomimetic ligand used for the purification of therapeutic proteins; phytoestrogens biochemical aspects and biological activities; increase in amino acids and ammonia during liquefaction of semen; the structure of inhibitor polymers and prebiosis with fermentable carbohydrates in pig nutrition.

Chapter 1 - G-quadruplex represents a group of unusual DNA secondary structures, based on Hoogsteen G-G pairing hydrophobic planar rings of four guanines. Many studies have reported that G-quadruplex structures appear to be involved in many important biological processes, such as DNA replication, gene expression and recombination, as well as cell transformation. The G-quadruplex has been shown to be specifically important in the structure of G-rich single-strand DNA at the end of telomeres. The presence of these structures in vivo has been demonstrated in a number of biological systems, ranging form ciliates to humans. For these reasons, in the last few years, the research of specific G-quadruplex ligands as potential drugs has gained major attention, in particular as telomeres targeting agents and telomerase inhibitors and thus potential anticancer drugs.

Many eminent scientists have been working in this exciting field, as demonstrated by the success of the First International Meeting on Quadruplex DNA, held in Louisville (KY, USA) in April 2007§. The specificity of the interactions between the studied ligands and different DNA structures (both duplex and quadruplex) is surely a main topic in this area, since it determines the different biological effects of these compounds. Several approaches have been used to explore this subject, and they will be briefly summarized in this chapter. Nevertheless, particular attention should be paid when comparing biological data from different labs on telomerase inhibition by these compounds. In fact, these data, which are very important from a pharmaceutical and medicinal chemistry point of view, are affected by many factors, such as the type of cells used to extract telomerase, the presence of proteins other than telomerase in the cell extract, the amount of cell extract and DNA substrates used, as well as other features that have not been fully standardized yet.

Chapter 2 - The successful commercialization of high-purity proteins and enzymes parallels the development and improvement of downstream processing technology. Biomimetic ligands find wide application in protein separation and purification, being the most promising affinity ligands of large-scale potential. In the present paper we evaluate the ability of the biomimetic ligand 4-amino-phenyl-oxanilic acid coupled to Sepharose CL-6B via 1,3,5-trichloro-2,4,6-triazine to bind and purify human monoclonal anti-HIV antibody 2F5 (mAb 2F5) from spiked maize extracts and recombinant influenza virus neuraminidase (NA) expressed in insect cells. Under selected conditions the affinity adsorbent exhibited high selectivity and purifying ability for mAb 2F5 and NA.

Chapter 3 - Phytoestrogens are natural compounds found in plants that possess estrogen-like activity. The major classes of phytoestrogens are isoflavones, lignans and coumestans. Isoflavones are structurally similar to mammalian estrogen and are found in large amounts in soybean and soy products. Lignans are found in flaxseed, whole grain cereals, teas and some vegetables, and coumestans are mainly present in clover and alfalfa sprouts. The major isoflavones are genistein, daidzein and glycitein. These compounds are hydrolysis products of glycosides in plants, termed genistin, daidzin and glycitin, which, after oral ingestion, are metabolized by intestinal flora. In lignans, the plant precursors are matairesinol and secoisolariciresinol, which are also metabolized by the intestinal microflora to enterolactone and enterodiol, respectively. Phytoestrogens have been largely used in the world and for a long time period, and it has been observed that the incidence of hormone-dependent diseases is reduced in countries, especially Asian countries, with a high dietary content of phytoestrogens. Animal models and clinical investigations have shown that phytoestrogens are protective in the prevention of hormone-dependent cancers, primarily breast and prostate cancers; cardiovascular disease; osteoporosis and to alleviate the symptoms of menopause. The estrogenic effects of the phytoestrogens are mediated through the estrogen receptors ERα and ERβ, which function as ligand-inducible transcription factors for genes involved in cell growth, proliferation, and differentiation. Many mechanisms of action have been identified for phytoestrogen prevention of diseases, including estrogenic/antiestrogenic activity, antiproliferation, induction of cell cycle arrest and apoptosis, antioxidation activity, induction of detoxification enzymes, regulation of the host immune system, and changes in cell signaling. Despite growing evidence of the health benefits of phytoestrogens, it is important to investigate further their mechanism of actions for a better understanding of the biological outcomes and possible adverse effects.

Chapter 4 - The metastable structure of a serpin a1-proteinase inhibitor (a1-PI) makes it prone to conformational changes as a result of certain mutations and under mild denaturing conditions. Polymers of several variants of a1-PI, e.g., the most clinically relevant Z variant, accumulate in the liver and are believed to be the underlying cause of the liver disease associated with a1-PI deficiency. Such polymers have also been found in the circulation and in the lung. The structure of these polymers, which is important for structure-based drug design, remains unknown. The historically first and generally accepted model of the a1-PI polymers, the loop-A sheet model, which assumes insertion of the reactive center loop (RCL) of one a1-PI molecule between the central strands of the A b-sheet of another molecule, has never been proven. In addition, this model is inconsistent with resonance energy transfer data obtained for heteropolymers formed from the Z and S variants. It is also difficult to envision how this model or even the more compact loop-A sheet model, deduced from fluorescence data obtained for the Z and S variant heteropolymers, can be compatible with electron microscopy (EM) images, which show apparently flexible polymer chains. Two other polymer models were deduced from the interactions seen between molecules in the crystals of two other serpins, antithrombin (AT) and plasminogen activator inhibitor-1 (PAI-1), which demonstrated the ability of the reactive center loop (RCL) to assume the b-strand conformation and to extend the C or A b-sheet in a neighboring molecule by formation of an additional edge strand. This became the ground for the loop-C sheet and strand s7A models of the polymer, although the interactions seen in crystals appear not to be stable in solution and, thus, appear not to reflect interactions existing in a1-PI polymers, which actually are stable. The focus of this paper is to review the observations used to support the proposed models and to revisit the wealth of literature data in search for unexplained observations and possible new interpretations. I also put forward the hypothesis that a fusion of b-sheets, rather than insertion of the reactive center loop into a b-sheet, may underlie the polymerization process of a1-PI. This hypothesis is an expansion of the recent head-to-head model proposed based on the polymerization of the a1-PI disulfide-linked dimer.

Chapter 5 - Fresh normal ejaculate clots by action of about 0.05 IU/ml thrombin on a fibrinogen-like compound, and subsequently lyses by action of about 0.1 IU/ml plasmin and about 300 % (of blood plasma norm) carboxypeptidase. Carboxypeptidases sequentially liberate amino acids, that were quantified in the present work: the most abundant amino acids in fresh seminal plasma are serine (6.09 mM), threonine (4.61 mM), lysine (3.45 mM), tyrosine (3.18 mM), glycine (2.79 mM), and arginine (2.29 mM). The concentration of the amino acids approximately doubles within 45 min (37°C), with an exceptional 12fold increase of glutamic acid. When compared to blood plasma, fresh seminal plasma contains 226fold gamma-glutamyl-transferase (γ-GT), 40fold ammonia, 21fold creatinine, and 18fold glutamate dehydrogenase (GLDH). Ammonia (NH3), generated by glutamate dehydrogenase (GLDH), together with arginine or creatinine might protect sperm cells from female defence cells. Ammonia and natural guanidine derivatives might be physiologic immunsuppressors.

Chapter 6 - Evolutionary processes over time have formed a symbiotic relationship between a well diverse and stable microbiota and the animal host. Berg [1] and Gaskins [2] have pointed out that bacterial cells outnumber animal cells by a factor of 10 and have a profound influence on nutritional, physiological, immunological, and protective processes in the host.

Essentially, gastrointestinal tracts of animals are sterile at birth. After the first introduction to microflora during birth, the GI tract gets colonized by microbes from mother and the environment, until a stable, diverse and complex microbiota exists. The successful development, stability, and functionality of this very complex ecosystem is dependent on the microbial community, host physiology and the diet (nutrient source for the ecosystem). Any alterations in these three pillars of the GIT microbial ecosystem, lead to changes in microbial activity with consequences on host physiology and performance.

This chapter is focused on the effects of prebiotics, essentially fermentable carbohydrates in the diet, on the gut microbial community, gut health, physiology and performance of the animal, and the environment in and around a pig farm. Author further discusses the candidature of fermentable carbohydrates as an alternative to in-feed antibiotics in pig production. Some recommendations for further research directions are also discussed.

Chapter 7 - Mycotoxins are toxic or carcinogenic compounds produced by fungi, including Aspergilllus and Fusarium molds that colonize seeds of maize, cotton, peanuts or tree nuts. Insect herbivory of plant tissue often enhances mold infection. Transgenic maize expressing the Bacillus thuringiensis (Bt) toxin produce ears with lower mycotoxin levels when the target insect pest is effectively controlled. However, Bt toxins are generally species specific and therefore new transgenic strategies for broad insect control need to be developed. The vast diversity of plant secondary biochemicals reflects the strategy of non-motile plants to defend themselves from pathogens and animal herbivores. Research in recent years revealed that many of these plant biochemicals are synthesized by a number of enzymes that are coordinately regulated at the gene level by transcription factors. Altering the expression of these transcription factors by genetic engineering opens the possibility that these secondary biochemicals may be synthesized in specific tissues to combat insect herbivory and reduce mycotoxin contamination. Alternatively, well studied secondary biochemical pathways may be modified or transferred from one plant species to another and tested for effective insect resistance. Transgenic manipulation of secondary metabolism must be carefully considered because redirection of plant energy reserves may hinder plant yield. The advent of genetic engineering and advances in plant biochemistry has opened up the exciting possibilities of utilizing the diversity of secondary biochemicals for protecting valuable crop commodities.

Versions of these chapters were also published in *International Journal of Medical and Biological Frontiers,* Volume 15, Numbers 3-4, 9-10, and 11-12, edited by Tsisana Shartava, published by Nova Science Publishers, Inc. They were submitted for appropriate modifications in an effort to encourage wider dissemination of research.

Ligands, Polymers and Amino Acids
Editor: Tsisana Shartava
ISBN: 978-1-61122-793-2

G-QUADRUPLEX LIGANDS: A SELECTIVE APPROACH TO TELOMERES LENGTH REGULATION, BOTH DIRECTLY AND THROUGH TELOMERASE INHIBITION

Marco Franceschin*[*1], Armandodoriano Bianco[1] and Maria Savino[2]
[1]Dipartimento di Chimica
[2]Dipartimento di Genetica e Biologia Molecolare,
Università di Roma "La Sapienza", Piazzale A. Moro 5, 00185 Roma, Italy

ABSTRACT

G-quadruplex represents a group of unusual DNA secondary structures, based on Hoogsteen G-G pairing hydrophobic planar rings of four guanines. Many studies have reported that G-quadruplex structures appear to be involved in many important biological processes, such as DNA replication, gene expression and recombination, as well as cell transformation. The G-quadruplex has been shown to be specifically important in the structure of G-rich single-strand DNA at the end of telomeres. The presence of these structures *in vivo* has been demonstrated in a number of biological systems, ranging form ciliates to humans. For these reasons, in the last few years, the research of specific G-quadruplex ligands as potential drugs has gained major attention, in particular as telomeres targeting agents and telomerase inhibitors and thus potential anticancer drugs.

Many eminent scientists have been working in this exciting field, as demonstrated by the success of the First International Meeting on Quadruplex DNA, held in Louisville (KY, USA) in April 2007[§]. The specificity of the interactions between the studied ligands and different DNA structures (both duplex and quadruplex) is surely a main topic in this area, since it determines the different biological effects of these compounds. Several approaches have been used to explore this subject, and they will be briefly summarized in this chapter. Nevertheless, particular attention should be paid when comparing biological data from different labs on telomerase inhibition by these compounds. In fact, these data, which are very important from a pharmaceutical and medicinal chemistry point of view, are affected by many factors, such as the type of cells used to extract telomerase, the presence of proteins other than telomerase in the cell extract, the amount of cell extract and DNA substrates used, as well as other features that have not been fully standardized yet.

[*] E-mail: marco.franceschin@uniroma1.it

INTRODUCTION

G-quadruplexes are a family of nucleic acid secondary structures stabilized by G-quartets that form in the presence of cations [1]. The level of interest in these structures has recently increased due to the hypothesis that G-quadruplex structures play roles in key biological processes [2]. Each quartet is composed of four guanines, held together by a cyclic arrangement of eight Hoogsteen hydrogen bonds (Figure 1). The presence of a central cation (typically potassium) helps to maintain the stability of the structure, which may be very stable under physiological conditions. Different G-quadruplex structures exist (Figure 2), depending on the orientation of the DNA strands and the syn/anti conformation of the guanines [3]. The human telomeric G-overhang can fold into several different intramolecular quadruplex structures that differ by the position of the adjacent loop regions [4]. Recently, the field of G-quadruplex study has been greatly enlarged since the in vivo existence of G-quadruplexes has been proposed for oncogene promoters and telomeres through the use of specific ligands [5].

Human cells telomeres protect chromosomal ends from fusion events; they range in size from 3 to 15 kb and are composed of tandem repeats of the sequence 5′-GGTTAG-3′ with a 3′ overhang of the G-strand which plays an important structural and functional role [6]. In human somatic cells, telomere length decreases with each cell division event [7]; reversion of this degradation by a telomere maintenance mechanism increases cellular replicative capacity. A telomere maintenance mechanism is provided by a specialized enzyme called telomerase [8]. In the majority of tumour cells (85–90%), this enzyme is overexpressed [9]. Even in normal stem cells, the level of telomerase activity is low or absent [10]. The catalytic reverse transcriptase subunit of the human enzyme, a protein called hTERT, uses the RNA subunit (hTR) as a template to add 5′-GGTTAG-3′ repeats to the ends of chromosomes, that is the G-rich strand of the telomere [11]. Telomerase is involved in telomere capping and in the DNA-damage response [12].

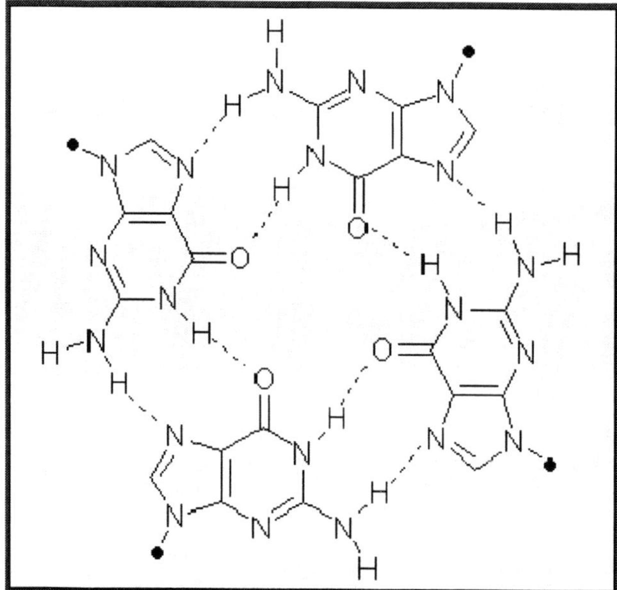

Figure 1. A G-tetrad showing Hoogsteen hydrogen bonds.

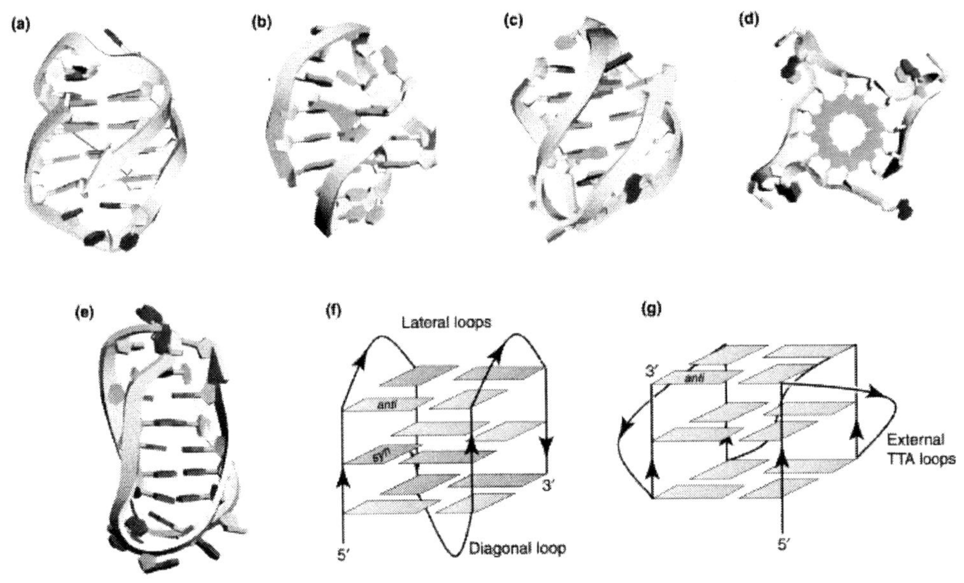

Figure 2. (a–e) Ribbon diagrams of various telomeric DNA quadruplex structures. d(G4T4G4) from Oxytricha (a); d(G4T4G3) (b); d[AG3(T2AG3)3] in Na+ solution (c) d[AG3(T2AG3)3] with K+ in the crystal (d); d(C2AT2C2AT2C2T3C2) in solution - the i-motif (e). (f,g) Representations of the folds in structures (c,d), respectively. [1] Reprinted from Curr. Opin. Struct. Biol. (13) Neidle S., Parkinson G.N. "The structure of telomeric DNA" 275-283, Copyright (2003), with permission from Elsevier.

Optimal telomerase activity requires an unfolded single-stranded substrate, because G-quadruplex formation directly inhibits telomerase elongation in vitro (Figure 3a) [13]. Therefore, ligands that selectively bind to and stabilize G-quadruplex structures may interfere with telomere conformation and telomere elongation [14]. For these reasons telomerase inhibitors have gained major attention as possible selective anticancer agents. In fact, when telomerase is activated in somatic cells, telomeres are maintained above the critical length that induces apoptosis and senescence, leading to uncontrolled proliferation: the inhibition of telomerase ensure correct telomere shortening and regular cellular senescence (Figure 4).

G-Quadruplex Interactive Compounds as Telomerase Inhibitors

A range of G-quadruplex ligands have been shown to bind quadruplexes in vitro (Figure 5) [15]. The types of molecules that are able to stabilize G-quadruplexes are characterized in most cases by an aromatic core, that favours stacking interactions with the G-tetrads, and basic side chains (positively charged in physiological conditions), which interact with the quadruple helix grooves, adopting a terminal stacking mode (Figure 6) [16]. These molecules recognize quadruplex DNA and are active in the telomere repeat amplification protocol (TRAP) assay [17-18]. Some of these molecules have also been shown to induce telomere shortening and/or telomere instability, triggering apoptosis and/or senescence programs, in various cell lines [19].

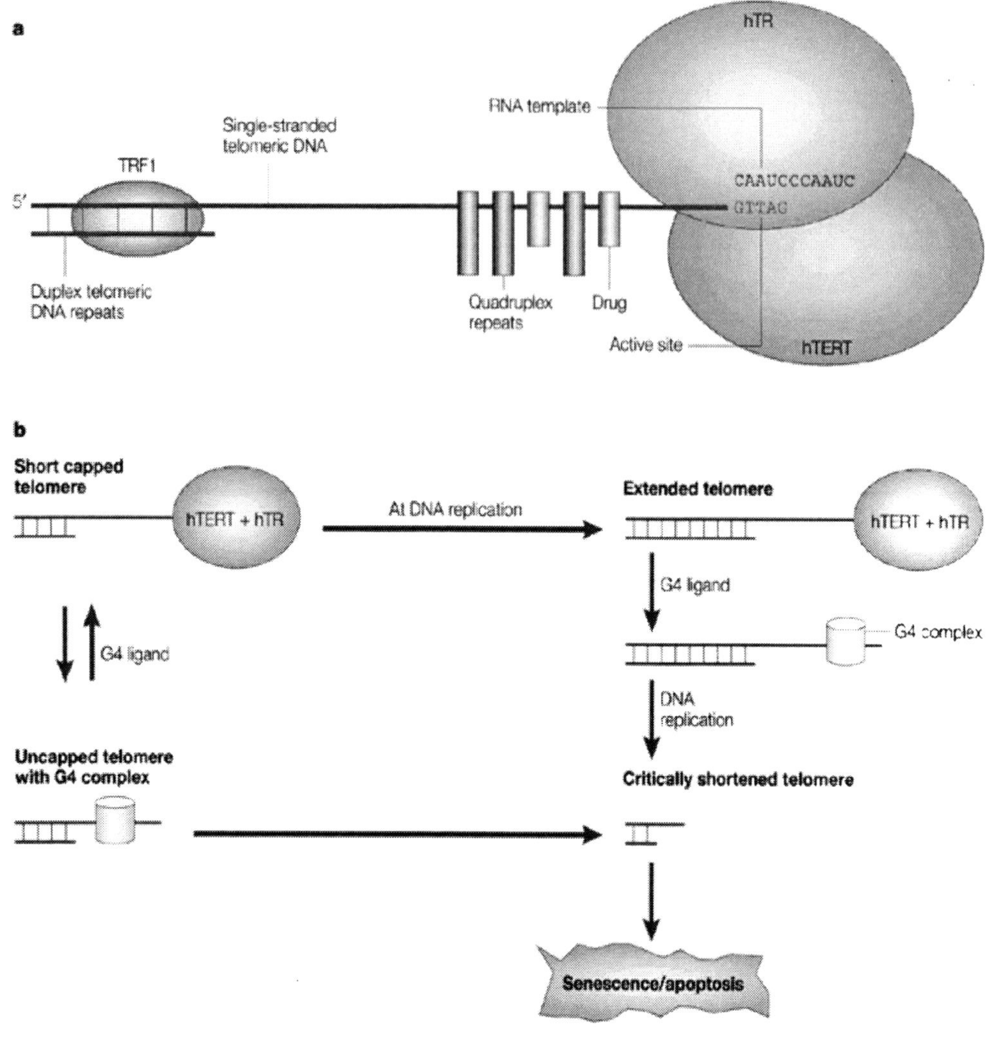

Figure 3. G-quadruplexes and telomerase inhibition. a) Schematic of telomerase inhibition being produced by folding of the 3′ end telomere primer strand into G-quadruplex units (which might assemble one onto another), and which are stabilized by ligands stacking onto the G-quartet ends. This folding then inhibits further hybridization of the primer with the hTR template from taking place during subsequent cycles of telomere extension. b) Schematic of the sequence of events that can take place after quadruplex-induced telomerase inhibition, with two pathways to selective cell death being shown, dependent on telomere length. [14] Reprinted by permission from Macmillan Publishers Ltd: Neidle, S.; Parkinson, G. "Telomere maintenance as a target for anticancer drug discovery" Nat. Rev. Drug Discov. 2002, 1, 383-393. Copyright 2002, Nature Publishing Group.

In order to inhibit telomerase, many molecules with these characteristics have been synthesized: anthraquinones [20-21] and 3,6-disubstituited acridines [22] are examples of the first generation of these inhibitors. Subsequently, tri-substituted acridines [23] porphyrins [24] and triazines [25] have been synthesised to improve the efficiency in inhibiting telomerase.

Natural compounds, such as berberine [26] and telomestatin [27], have also been shown to inhibit human telomerase, the latter with high efficiency. Perylene diimides present optimal features to interact with the G-quadruplex [28] and show a good ability to induce different G-quadruplex structures and to inhibit telomerase [29], depending on the side chains basicity and length [30-31]. Our research group has recently proposed a general synthetic protocol, preparing a series of new highly hydrosoluble perylene [32-33] and coronene [34] derivatives with two up to four side chains having various basic substituents. All of them exhibit an enhanced ability to interact with G-quadruplex structures and to inhibit human telomerase, with respect to the previously reported perylene derivatives.

Our study was carried out by investigating the ability of the synthesized compounds to induce inter and intramolecular G-quadruplex structures by polyacrylamide gel electrophoresis (PAGE) and to inhibit telomerase in a modified TRAP assay. The two properties appear to be satisfactorily correlated.

More recently, several macrocycles, based on quinacridine [35], oxazole [36] and quinoline [37] moieties, have been proposed and studied as G-quadruplex ligands. It is worth noting that, due to their direct interaction with telomeres, G-quadruplex ligands have shown more rapid and specific effects than those expected for simple telomerase inhibitors (Figure 4b), probably by perturbing the equilibrium of capped/uncapped telomeres (Figure 3b) [38-39].

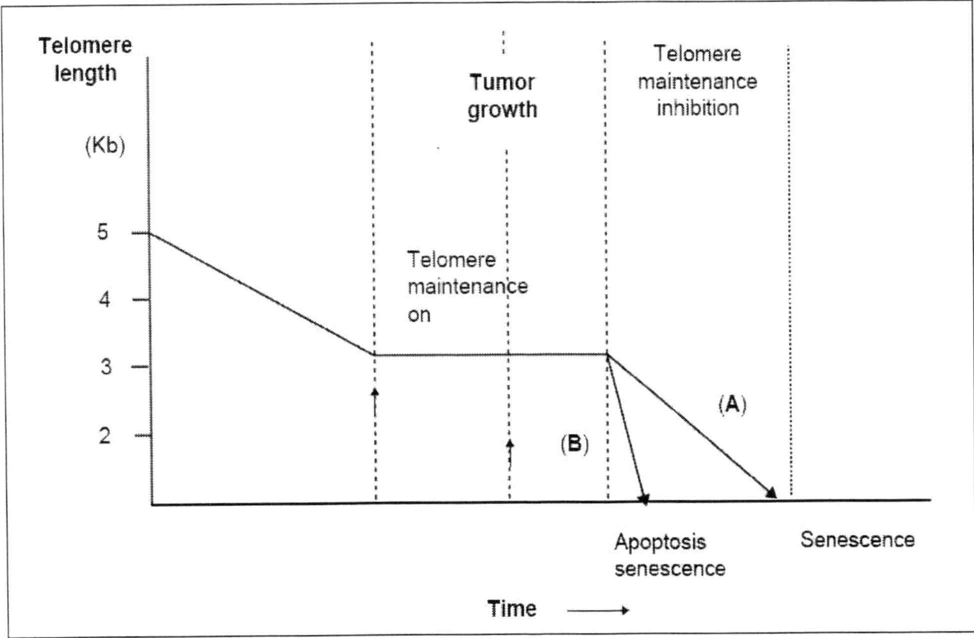

Figure 4. The progression of telomeres from normal via telomerase activation through to telomerase inhibition and senescence. (A) on the basis of the classic model of telomere maintenance, (B) on the basis of non-classical, actual behavior of quadruplex inhibitors with shortened telomeres. [15] Reproduced, with permission from The Thomson Corporation from Incles CM, Schultes CM, Neidle S: Telomerase inhibitors in cancer therapy: Current status and future directions. Current Opinion in Investigational Drugs (2003) 4(6):675-685. Copyright 2003, The Thomson Corporation.

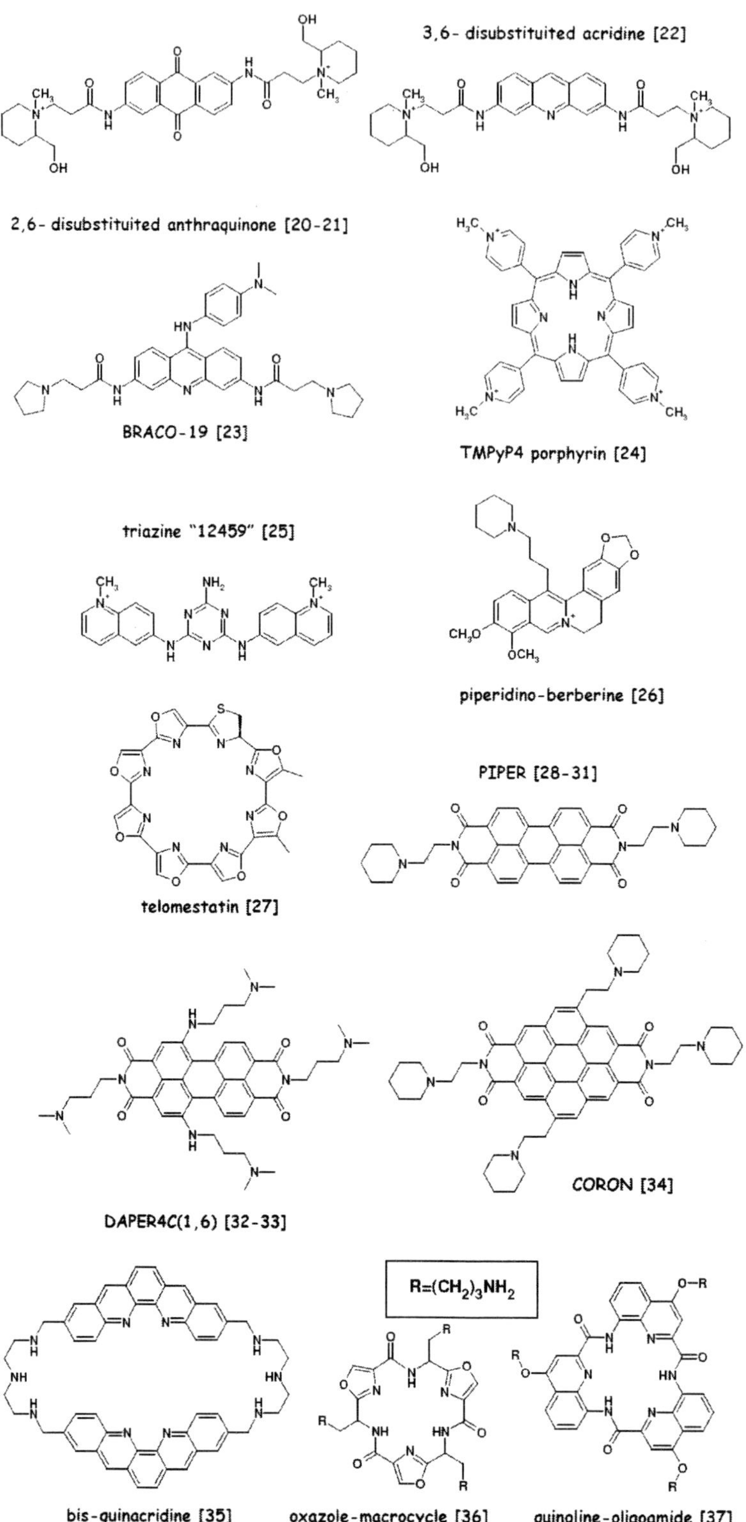

Figure 5. Structures of known G-quadruplex ligands, in square brackets the cited reference(s).

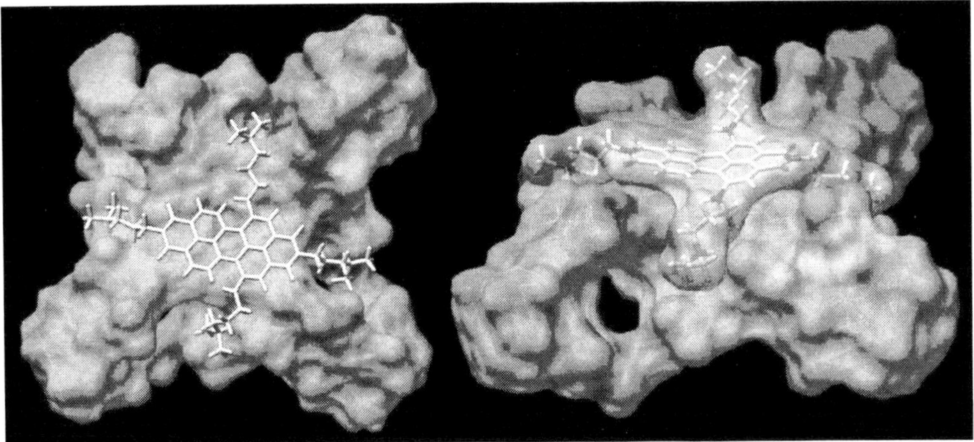

Figure 6. Representation of a complex between a perylene derivative [DAPER4C(1,6)] and a monomeric G-quadruplex (blue surface, K+ ions are violet spheres) obtained by simulated annealing [33]. The ligand is in yellow as sticks (left) and as atom type colored sticks and yellow surfaces (right, under a different perspective).

STUDY OF THE INTERACTIONS OF THE LIGANDS WITH QUADRUPLEX AND DUPLEX DNA

Detailed studies on drug-quadruplex complexes are essential to understand quadruplex recognition and address drug design [40]. Several techniques can be used to this aim, including absorption spectroscopy [41-42], circular dichroism [43-44], emission spectroscopy [45], calorimetry [46], nuclear magnetic resonance (NMR) [47-49], surface plasmon resonance (SPR) [50-51], mass spectrometry [52], X-ray diffraction [53-54] and competition dialysis [55]. In particular, fluorescence-based melting assays for studying quadruplex ligands have been developed, and specifically Fluorescence Resonance Energy Transfer (FRET) assays have been widely used to study the thermal stabilization of preformed G-quadruplex structures upon binding of different ligands [56-57]. Once two probes forming a donor-acceptor system are linked at the end of G-quadruplex forming sequences, FRET assays represent a powerful tool for the study of G-quadruplex ligands, due to the possibility to perform a high number of simultaneous measurements in different conditions (including ionic strength and various drug concentrations).

Mass spectrometry is a novel tool for studying biomolecular structures and non-covalent interactions; it can provide data on the functional properties of biomacromolecules complementary to that obtained from more traditional techniques, and specifically electrospray ionization-mass spectrometry (ESI-MS) is very useful in the analysis of non-covalent complexes between nucleic acids and small molecules [58], since under certain conditions it allows the transfer of non-covalently bound complexes into the gas phase without the disruption of the complex and therefore the mass spectrometric determination of their modes and energies of interaction [59]. Our group has recently evaluated the ability of the newly synthesized ligands to interact with duplex and quadruplex DNA structures, analysing the complexes formed with model oligonucleotides via ESI-MS and comparing the obtained results with those obtained by other techniques and from the biological tests.

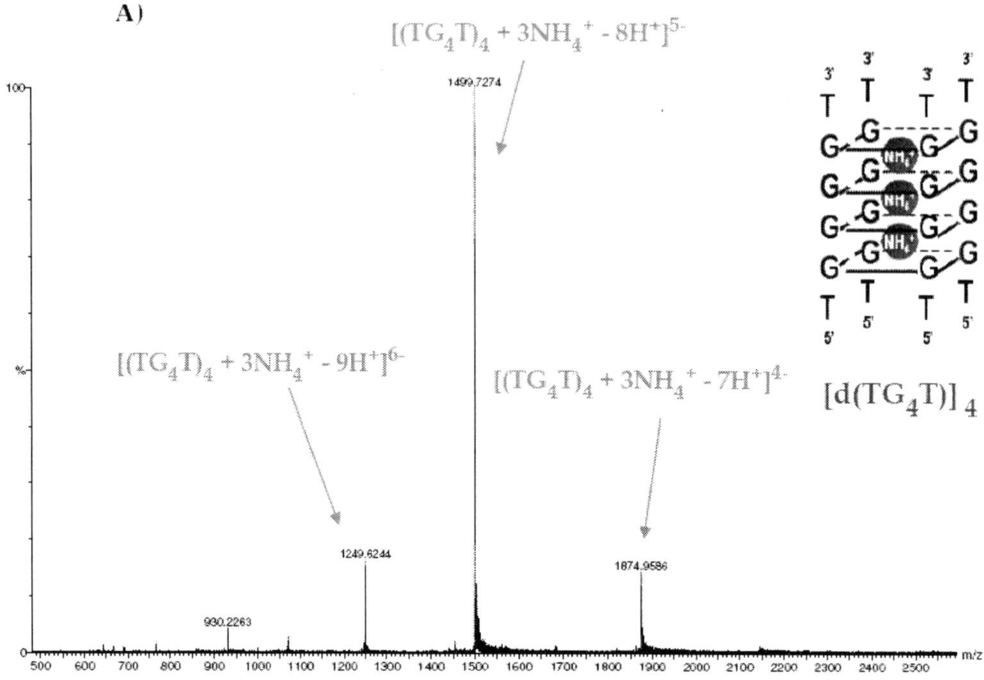

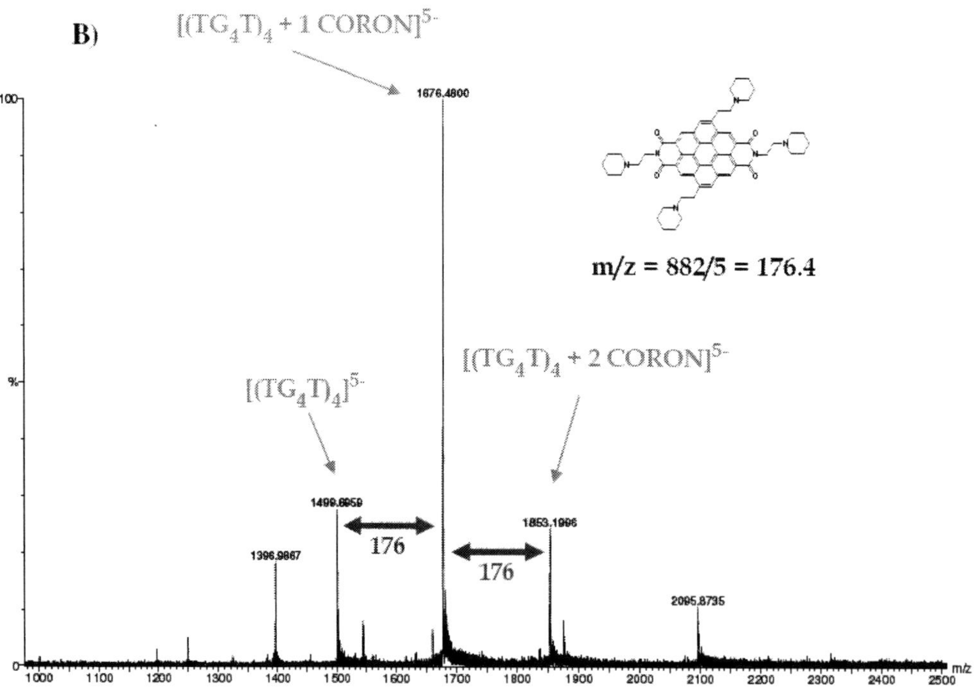

Figure 7. ESI mass spectra for complexes containing [d(TG4T)]4 alone (A) and with CORON (B). Labels and arrows indicate the main ions identified: in the second panel counterions are omitted. Since ions are multi-charged (5-), the complexes containing one or more ligand molecules differ only of 1/5 of the ligand mass. Thanks are due to Antonello Alvino for these spectra.

In particular, we found this technique very useful to determine the quadruplex vs duplex selectivity of the synthesized ligands and the determination of the stoichiometry of the complexes formed with DNA and the relative binding constants (unpublished results). As an example, in figure 7, ESI mass spectra for complexes between $[d(TG_4T)]_4$ and CORON are reported, showing drug/quadruplex complexes at both 1:1 and, to a minor extent, 2:1 ratios. These data confirm the existence of two binding sites, according to the terminal stacking model (Figure 6).

The selectivity between quadruplex and duplex DNA is surely a highly relevant topic and could be related to the specificity of the biological activity of these compounds [60]. This topic can be investigated with most of the previously mentioned techniques [61], choosing an appropriate model for duplex DNA: Dickerson dodecamer or genomic DNA (i.e. calf thymus DNA), depending on the technique used. It is particularly interesting the possibility to perform the same experiments described before in the simultaneous presence of duplex and quadruplex DNA [62].

From a kinetic point of view, the study of the formation of the G-quadruplex upon addition of specific ligands can be efficiently studied by polyacrylamide gel electrophoresis (PAGE) [28]. In fact, it is possible to study the induction of different G-quadruplex structures by various ligands by means of this technique [30-34]. Using oligonucleotides with different G-rich sequences and considering the standard mobility of well known monomeric, dimeric and tetrameric G-quadruplexes, it is possible to assign the bands formed in the presence of various ligands to complexes corresponding to different quadruplex conformations. Many of the G-quadruplex ligands can act as molecular chaperones, assisting the G-quadruplex folding of single-strand DNA and reducing the time needed for its formation, other than stabilizing the final structure (Figure 8) [63].

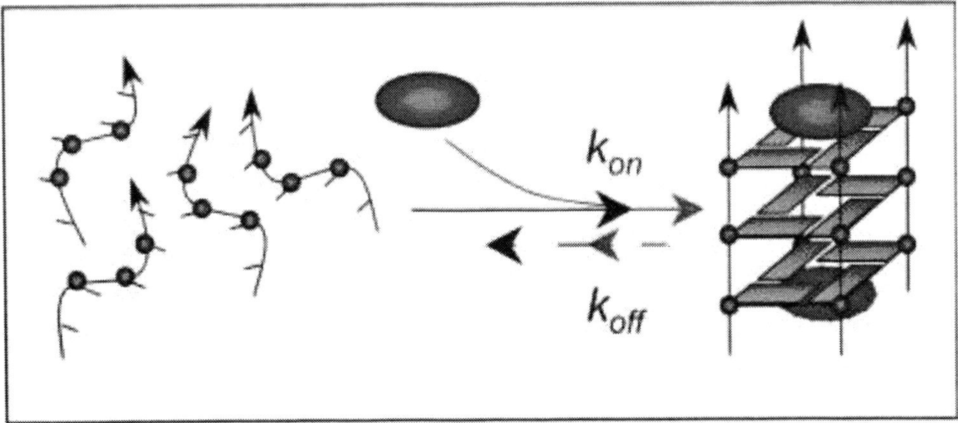

Figure 8. Tetramolecular G-quadruplex formation/dissociation. This scheme simply presents the initial and final states of the reaction and assumes terminal stacking of the ligand; it does not claim to reflect the pathway of association. Are quadruplex ligands (red ovals) simply able to lock the preformed structure (i.e. increase the lifetime of the structure by lowering its dissociation constant koff), or do ligands actively promote the formation of the complex and act as quadruplex chaperones by increasing its association constant kon? [63] Reprinted from Nucleic Acids Res. (35) De Cian A., Mergny J.L. "Quadruplex ligands may act as molecular chaperones for tetramolecular quadruplex formation" 2483-2493, Copyright (2007), with permission from Oxford University Press and the authors.

The self-aggregation of the synthesized compounds in aqueous solution must be carefully considered in the model of the interaction between the different ligands and G-quadruplex DNA. In fact, self-association is reported to favour the specific recognition of the G-quadruplex with respect to duplex DNA [64-66]. Nevertheless, it has been recently shown by Palumbo and coworkers [29] that strong drug self-aggregation is related to a lower telomerase inhibition and weaker interactions with the G-quadruplex, suggesting that the higher selectivity for G-quadruplex arrangements upon aggregation is due to a reduced binding efficiency to duplex and single stranded DNA rather than to a higher affinity for G-quartets.

Study of the Biological Activity (TRAP Assay)

The Telomeric Repeat Amplification Protocol (TRAP) assay is commonly used to evaluate telomerase activity in tissues or cell extracts and also to determine the inhibitory properties of small molecules against telomerase [67]. It consists of two different reactions: the first one is the telomerase elongation of the substrate, while the second one is the amplification of the products by polymerase chain reaction (PCR). The original method was improved by the addition of a tag sequence at the 5'-end of the reverse primer and the use of an internal control that coamplifies with the telomerase products and allows the detection of Taq inhibitors [68]. Both TS and TSG4 oligonucleotides can be used as primers for the elongation by telomerase (Figure 9). The first one does not contain the guanine repeats characteristic of the telomeric sequence, while TSG4 is designed to mimic the GGG tracts of the telomeric sequence, but having the loop sequence inverted: this feature decreases the thermodynamic stability of the intramolecular G-quadruplex with respect to that formed by the human telomeric sequence. The elongation products, in the presence of different concentrations of putative inhibitors, are then amplified by PCR and finally visualized by radioactivity or fluorescence [69]. In the case of G-quadruplex ligands the primer TSG4, presenting four repeats rich in guanine, looks particularly suitable for the study of the biological activity of these ligands [31, 69].

Several factors can influence the results of the TRAP assay and thus the quantification of the inhibitory activity of different compounds. First of all the concentration of the primer: it is clearly a fundamental parameter, but surprisingly in many kits that are now commercially available it is not indicated. Another very important aspect to be considered is that a cell extract is used rather than the purified enzyme, so that the type of cells used to extract telomerase, the presence of proteins other than telomerase in the cell extract and the quantity of cell extract used represent important parameters which have not been fully standardized yet. In this sense, the possibility to extract telomerase from cells overexpressing the enzyme is surely of great interest [70].

For these reasons, particular attention should be paid when comparing biological data on telomerase inhibition by various compounds coming from different labs, even though it is tempting to simply compare the IC_{50} values (i.e. the concentration of an inhibitor required to obtain 50% inhibition of telomerase activity *in vitro*), as made in many reviews on telomerase inhibitors [15, 71]. As an example of this, the case of the well known perylene derivative PIPER can be considered: many studies from different groups agree with an IC_{50} for this compound between 10 and 20 μM [30-31, 47] but recently Palumbo and coworkers reported an IC_{50} of 0.2 μM [29].

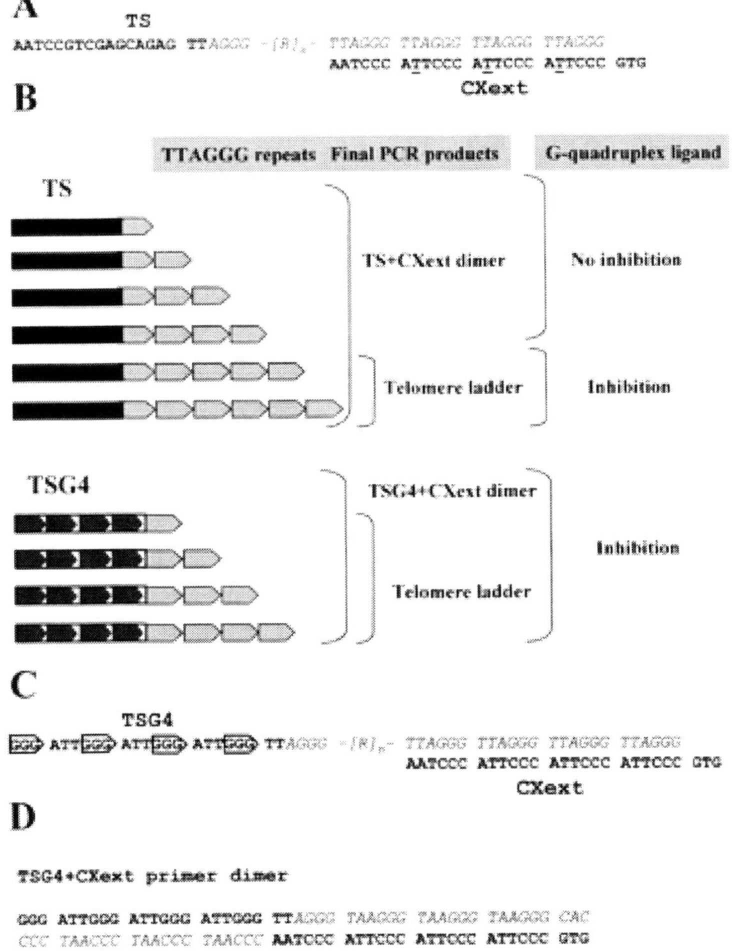

Figure 9. A, sequence of TS primer. Telomerase elongates the TS primer, adding TTAGGG repeats. During PCR, CXext hybridizes to the 5' end of the repeats in subsequent PCR steps; TS and CXext primers are shown in bold. Mismatched nucleotides are underlined. [R]n corresponds to n repetitions of TTAGGG. B, schematic representation of the telomerase products (TTAGGG repeats) extended from: the TS primer in TRAP (upper part) or from the TSG4 primer in TRAP-G4 (lower part). When 0–4 TTAGGG repeats are added to TS by telomerase, final PCR products correspond to TS+CXext dimer; a G-quadruplex ligand should not have an inhibitory effect on their formation. When >4 TTAGGG repeats are added by telomerase, final PCR products correspond to TS+CXext dimer and telomere ladder. A G-quadruplex ligand should inhibit telomere ladder formation. In the presence of TSG4, a quadruplex ligand should inhibit both TSG4+CXext dimer and telomere ladder formation. C, sequence of TSG4 primer. Telomerase elongates the TSG4 primer, adding TTAGGG repeats. CXext hybridize to the 5'-end of the repeats in subsequent PCR steps. TSG4 and CXext primers are shown in bold. Mismatched nucleotides are underlined. [R]n corresponds to n repetitions of TTAGGG. D, minimal overlap leading to TSG4+CXext primer dimer formation that yields elongated TSG4 with four added repeats. [69]. Reprinted from Cancer Res. (62) Gomez D., Mergny J.L., Riou J.F. "Detection of telomerase inhibitors based on g-quadruplex ligands by a modified telomeric repeat amplification protocol assay" 3365-3368, Copyright (2002), with permission from the Copyright Clearance Center of the American Association for Cancer Research and the rightsholder.

Moreover, the inhibition of Taq polymerase by the same ligands should be carefully considered in the PCR step. In order to overcome this problem, an internal standard (IS) is usually inserted in the PCR reaction, to control the correct and complete amplification [69]. Nevertheless, since the IS oligonucleotide sequence is not rich in guanine, it has been recently shown that it could be inappropriate to evaluate Taq polymerase inhibition by quadruplex ligands [72]. In fact, due to their strong interaction with G-rich sequences, G-quadruplex specific ligands could inhibit the polymerase activity on the G-rich product of elongation, even when PCR controls are included and not affected. This means that the standard TRAP assay could be inappropriate for the quantification of telomerase inhibition in this case and the inhibitory effect on human telomerase of many quadruplex ligands could have been overestimated. Indeed, several G-quadruplex ligands have been studied in a Taq Polymerase Stop Assay [50], with the aim to evaluate their ability to inhibit the transcription of oncogenes like c-myc, presenting G-rich sequences in their promoters [24].

In order to overcome this problem, two different strategies are possible. One is to perform a "direct assay" [72], that is without the PCR amplification: this requires telomerase-enriched cell extracts [70] and a much higher quantity (≈10-fold) of radioactive material than a typical TRAP assay. The second possibility to circumvent these difficulties is to remove the inhibitor after the telomerase extension step (i.e., before PCR): this can be obtained with solvent extraction and/or precipitation [30-31]. Sometimes it has been argued about the reliability of this approach, since it is laborious and could result in poor reproducibility, due to the difficulties in obtaining a quantitative rescue of DNA. Nevertheless, if the internal standard (IS) is used to evaluate the efficiency of DNA rescue, other than the PCR activity, and to standardize the quantification of elongation products in each lane, we have experienced a good reproducibility of the experiments and reliable results [33-34].

Conclusion

G-quadruplex interactive compounds surely represent a very interesting class of bioactive molecules, not only as possible telomerase inhibitors, but as specific agents acting directly on telomere structure as well as on other G-rich regions of the genome, especially the promoters of several oncogenes. The interest towards these compounds is demonstrated by the impressive amount of papers on their structural and biological studies by scientists throughout the world. Many aspects still need to be fully clarified, above all the structural selectivity for different G-quadruplex structures and for quadruplex with respect to duplex DNA. Several studies on the biological effects of these ligands have been performing in many different laboratories and could lead to interesting and unexpected results. Even though, as widely explained in this chapter, it is necessary to pay attention to compare results coming from different labs, these molecules surely deserve further investigation, due to their high potentiality as new selective anticancer drugs. It is worth to underline that, in this interdisciplinary field, the collaboration between research groups having different expertises is particularly significant.

REFERENCES

[1] Neidle, S.; Parkinson, G.N. "The structure of telomeric DNA" *Curr. Opin. Struct. Biol.* 2003, *13*, 275-283.

[2] Paeschke, K.; Simonsson, T.; Postberg, J.; Rhodes, D.; Lipps, H.J. "Telomere end-binding proteins control the formation of G-quadruplex DNA structures in vivo" *Nat. Struct. Mol. Biol.* 2005, *12*, 847-854.

[3] Burge, S.; Parkinson, G.N.; Hazel, P.; Todd, A.K.; Neidle, S. "Quadruplex DNA: sequence, topology and structure" *Nucleic Acids Res.* 2006, *34*, 5402-5415.

[4] Luu, K.N.; Phan, A.T.; Kuryavyi, V.; Lacroix, L.; Patel, D.J. "Structure of the human telomere in K^+ solution: an intramolecular (3 + 1) G-quadruplex scaffold" *J. Am. Chem. Soc.* 2006, *128*, 9963-9970.

[5] Granotier, C.; Pennarun, G.; Riou, L.; Hoffschir, F.; Gauthier, L.R.; De Cian, A.; Gomez, D.; Mandine, E.; Riou, J.F.; Mergny, J.L.; Mailliet, P.; Dutrillaux, B.; Boussin, F.D. "Preferential binding of a G-quadruplex ligand to human chromosome ends" *Nucleic Acids Res.* 2005, *33*, 4182-4190.

[6] Meyne, J.; Ratliff, R.L.; Moyzis, R.K. "Conservation of the human telomere sequence (TTAGGG)n among vertebrates" *Proc. Natl. Acad. Sci. USA* 1989, *86*, 7049-7053.

[7] Harley, C.B.; Futcher A.B.; Greider C.W. "Telomeres shorten during ageing of human fibroblasts" *Nature* 1990, *345*, 458-460.

[8] Shay, J.W.; Wright, W.E. "Telomerase: a target for cancer therapeutics" *Cancer cell* 2002, *2*, 257-265.

[9] Kim, N.W.; Piatyszek, M.A.; Prowse, K.R.; Harley, C.B.; West, M.D.; Ho, P.L.; Coviello, G.M.; Wright, W.E.; Weinrich, S.L.; Shay, J.W. "Specific association of human telomerase activity with immortal cells and cancer" *Science* 1994, *266*, 2011-2015.

[10] Hiyama, E.; Hiyama, K. "Telomere and telomerase in stem cells" *Br. J. Cancer* 2007, *96*, 1020.

[11] Theimer C.A.; Feigon, J. "Structure and function of telomerase RNA" *Curr. Opin. Struct. Biol.* 2006, *16*, 307-318.

[12] Autexier, C.; Lue, N.F. "The structure and function of telomerase reverse transcriptase" *Annu. Rev. Biochem.* 2006, *75*, 493-517.

[13] Zahler, A.M.; Williamson, J.R.; Cech, T.R.; Prescott, D.M. "Inhibition of telomerase by G-quartet DNA structures" *Nature* 1991, *350*, 718-720.

[14] Neidle, S.; Parkinson, G. "Telomere maintenance as a target for anticancer drug discovery" *Nat. Rev. Drug Discov.* 2002, *1*, 383-393.

[15] Incles, C.M.; Schultes, C.M.; Neidle, S. "Telomerase inhibitors in cancer therapy: current status and future directions" *Curr. Opin. Investig. Drugs* 2003, *4*, 675-685.

[16] Neidle, S.; Harrison, R.J.; Reszka, A.P.; Read, M.A. "Structure-activity relationships among guanine-quadruplex telomerase inhibitors" *Pharmacol. Ther.* 2000, *85*, 133-139.

[17] Haider, S.M.; Parkinson, G.N.; Neidle, S. "Structure of a G-quadruplex-ligand complex" *J. Mol. Biol.* 2003, *326*, 117-125.

[18] Clark, G.R.; Pytel, P.D.; Squire, C.J.; Neidle, S. "Structure of the first parallel DNA quadruplex-drug complex" *J. Am. Chem. Soc.* 2003, *125*, 4066-4067.

[19] Salvati, E.; Leonetti, C.; Rizzo, A.; Scarsella, M.; Mottolese, M.; Galati, R.; Sperduti, I.; Stevens, M.F.; D'Incalci, M.; Blasco, M.; Chiorino, G.; Bauwens, S.; Horard, B.; Gilson, E.; Stoppacciaro, A.; Zupi, G.; Biroccio, A. "Telomere damage induced by the G-quadruplex ligand RHPS4 has an antitumor effect" *J. Clin. Invest.* 2007, *117*, 3236-3247.

[20] Sun, D.; Thompson, B.; Cathers, B.E.; Salazar, M.; Kerwin, S.M.; Trent, J.O.; Jenkins, T.C.; Neidle, S.; Hurley, L.H. "Inhibition of human telomerase by a G-quadruplex-interactive compound" *J. Med. Chem.* 1997, *40*, 2113-2116.

[21] Huang, H.S.; Chou, C.L.; Guo, C.L.; Yuan, C.L.; Lu, Y.C.; Shieh, F.Y.; Lin, J.J. "Human telomerase inhibition and cytotoxicity of regioisomeric disubstituted amidoanthraquinones and aminoanthraquinones" *Bioorg. Med. Chem.* 2005, *13*, 1435-1444.

[22] Harrison, R.J.; Gowan, S.M.; Kelland, L.R.; Neidle, S. "Human telomerase inhibition by substituted acridine derivatives" *Bioorg. Med. Chem. Lett.* 1999, *9*, 2463-2468.

[23] Read, M.A.; Harrison, R.J.; Romagnoli, B.; Tanious, F.A.; Gowan, S.H.; Reszka, A.P.; Wilson, W.D.; Kelland, L.R.; Neidle, S. "Structure-based design of selective and potent G quadruplex-mediated telomerase inhibitors" *Proc. Natl. Acad. Sci. USA* 2001, *98*, 4844-4849.

[24] Seenisamy, J.; Bashyam, S.; Gokhale, V.; Vankayalapati, H.; Sun, D.; Siddiqui-Jain, A.; Streiner, N.; Shin-Ya K.; White, E.; Whilson, W.D.; Hurley, L.H. "Design and synthesis of an expanded porphyrin that has selectivity for the c-MYC G-quadruplex structure" *J. Am. Chem. Soc.* 2005, *127*, 2944-2959.

[25] Gomez, D.; Aouali, N.; Renaud, A.; Douarre, C.; Shin-Ya, K.; Tazi, J.; Martinez, S.; Trentesaux, C.; Morjani, H.; Riou, J.F. "Resistance to senescence induction and telomere shortening by a G-quadruplex ligand inhibitor of telomerase" *Cancer Res.* 2003, *63*, 6149-6153.

[26] Franceschin M., Rossetti L., D'Ambrosio A., Schirripa S., Bianco A., Ortaggi G., Savino M., Schultes C., Neidle S. "Natural and synthetic G-quadruplex interactive berberine derivatives" *Bioorg. Med. Chem. Lett.* 2006, *16*, 1707-1711.

[27] Kim, M.Y.; Vankayalapati, H.; Shin-ya, K.; Wierzba, K.; Hurley, L.H. "Telomestatin, a potent telomerase inhibitor that interacts quite specifically with the human telomeric intramolecular g-quadruplex" *J. Am.Chem. Soc.* 2001, *124*, 2098-2099.

[28] Han, H.; Cliff, C.L.; Hurley, L.H. "Accelerated assembly of G-quadruplex structures by a small molecule" *Biochemistry* 1999, *38*, 6981-6986.

[29] Sissi, C.; Lucatello, L.; Paul Krapcho, A.; Maloney, D.J.; Boxer, M.B.; Camarasa, M.V.; Pezzoni, G.; Menta, E.; Palumbo, M. "Tri-, tetra- and heptacyclic perylene analogues as new potential antineoplastic agents based on DNA telomerase inhibition" *Bioorg. Med. Chem.* 2007, *15*, 555-562.

[30] Rossetti, L.; Franceschin, M.; Bianco, A.; Ortaggi, G.; Savino, M. "Perylene diimides with different side chains are selective in inducing different G-quadruplex DNA structures and in inhibiting telomerase" *Bioorg. Med. Chem. Lett.* 2002, *12*, 2527-2533.

[31] Rossetti, L.; Franceschin, M.; Schirripa, S.; Bianco, A.; Ortaggi, G.; Savino, M. "Selective interactions of perylene derivatives having different side chains with inter- and intramolecular G-quadruplex DNA structures. A correlation with telomerase inhibition" *Bioorg. Med. Chem. Lett.* 2005, *15*, 413-420.

[32] Alvino, A.; Franceschin, M.; Cefaro, C.; Borioni, S.; Ortaggi, G.; Bianco, A. "Synthesis and spectroscopic properties of highly water-soluble perylene derivatives" *Tetrahedron* 2007, *63,* 7858-7865.

[33] Franceschin, M.; Pascucci, E.; Alvino, A.; D'Ambrosio D.; Bianco A.; Ortaggi, G.; Savino, M. "New highly hydrosoluble and not self-aggregated perylene derivatives with three and four polar side-chains as G-quadruplex telomere targeting agents and telomerase inhibitors" *Bioorg. Med. Chem. Lett.* 2007, *17*, 2515-2522.

[34] Franceschin, M.; Alvino, A.; Casagrande, V.; Mauriello, C.; Pascucci, E.; Savino, M.; Ortaggi, G.; Bianco A. "Specific interactions with intra- and intermolecular G-quadruplex DNA structures by hydrosoluble coronene derivatives: a new class of telomerase inhibitors" *Bioorg. Med. Chem.* 2007, *15*, 1848-1858.

[35] Teulade-Fichou, M.P.; Carrasco, C.; Guittat, L.; Bailly, C.; Alberti, P.; Mergny, J.L.; David, A.; Lehn, J.M.; Wilson, W.D. "Selective recognition of G-quadruplex telomeric DNA by a bis(quinacridine) macrocycle" *J. Am. Chem. Soc.* 2003, *125*, 4732-4740.

[36] Jantos, K.; Rodriguez, R.; Ladame, S.; Shirude, P.S.; Balasubramanian, S. "Oxazole-based peptide macrocycles: a new class of G-quadruplex binding ligands" *J. Am. Chem. Soc.* 2006, *128*, 13662-13663.

[37] Shirude, P.S.; Gillies, E.R.; Ladame, S.; Godde, F.; Shin-Ya, K.; Huc, I.; Balasubramanian, S. "Macrocyclic and helical oligoamides as a new class of G-quadruplex ligands" *J. Am. Chem. Soc.* 2007, *129*, 11890-11891.

[38] Neidle, S.; Read, M.A. "G-quadruplexes as therapeutic targets" *Biopolymers* 2000-2001, *56*, 195-208

[39] Riou, J.F. "G-quadruplex interacting agents targeting the telomeric G-overhang are more than simple telomerase inhibitors" *Curr. Med. Chem. Anticancer Agents* 2004, *4*, 439-443.

[40] Pagano, B.; Giancola, C. "Energetics of quadruplex-drug recognition in anticancer therapy" *Curr. Cancer Drug Targets* 2007, *7*, 520-540.

[41] Wei, C.; Jia, G.; Yuan, J.; Feng, Z.; Li, C. "A spectroscopic study on the interactions of porphyrin with G-quadruplex DNAs" *Biochemistry* 2006, *45*, 6681-6691.

[42] Sun, H.; Tang, Y.; Xiang, J.; Xu, G.; Zhang, Y.; Zhang, H.; Xu, L. "Spectroscopic studies of the interaction between quercetin and G-quadruplex DNA" *Bioorg. Med. Chem. Lett.* 2006, *16*, 3586-3589.

[43] Yamashita, T.; Uno, T.; Ishikawa, Y. "Stabilization of guanine quadruplex DNA by the binding of porphyrins with cationic side arms" *Bioorg. Med. Chem.* 2005, *13*, 2423-2430.

[44] Paramasivan, S.; Rujan, I.; Bolton, P.H. "Circular dichroism of quadruplex DNAs: Applications to structure, cation effects and ligand binding" *Methods* 2007, *43*, 324-331.

[45] Keating, L.R.; Szalai, V.A. "Parallel-stranded guanine quadruplex interactions with a copper cationic porphyrin" *Biochemistry* 2004, *43*, 15891-15900.

[46] Erra, E.; Petraccone, L.; Esposito, V.; Randazzo, A.; Mayol, L.; Ladbury, J.; Barone, G.; Giancola, C. "Interaction of porphyrin with G-quadruplex structures" *Nucleosides Nucleotides Nucleic Acids* 2005, *24*, 753-756.

[47] Fedoroff, O.Y.; Salazar, M.; Han, H.; Chemeris, V.V.; Kerwin, S.M.; Hurley, L.H. "NMR-Based model of a telomerase-inhibiting compound bound to G-quadruplex DNA" *Biochemistry* 1998, 37, 12367-12374.

[48] Hounsou, C.; Guittat, L.; Monchaud, D.; Jourdan, M.; Saettel, N.; Mergny, J.L.; Teulade-Fichou, M.P. "G-Quadruplex Recognition by Quinacridines: a SAR, NMR, and Biological Study" *ChemMedChem* 2007, *2*, 655-666.

[49] Randazzo, A.; Galeone, A.; Esposito, V.; Varra, M.; Mayol, L. "Interaction of distamycin A and netropsin with quadruplex and duplex structures: a comparative 1H-NMR study" *Nucleosides Nucleotides Nucleic Acids* 2002, *21*, 535-545.

[50] Rezler, E.M.; Seenisamy, J.; Bashyam, S.; Kim, M.Y.; White, E.; Wilson, W.D.; Hurley, L.H. "Telomestatin and diseleno sapphyrin bind selectively to two different forms of the human telomeric G-quadruplex structure" *J. Am. Chem. Soc.* 2005, *127*, 9439-9447.

[51] Redman, J.E. "Surface plasmon resonance for probing quadruplex folding and interactions with proteins and small molecules" *Methods* 2007, *43*, 302-312.

[52] Mazzitelli, C.L.; Brodbelt, J.S.; Kern, J.T.; Rodriguez, M.; Kerwin, S.M. "Evaluation of binding of perylene diimide and benzannulated perylene diimide ligands to DNA by electrospray ionization mass spectrometry" *J. Am. Soc. Mass Spectrom.* 2006, *17*, 593-604.

[53] Campbell, N.H.; Parkinson, G.N. "Crystallographic studies of quadruplex nucleic acids" *Methods* 2007, *43*, 252-263.

[54] Parkinson, G.N.; Ghosh, R.; Neidle, S. "Structural basis for binding of porphyrin to human telomeres" *Biochemistry* 2007, *46*, 2390-2397.

[55] Ragazzon, P.; Chaires, J.B. "Use of competition dialysis in the discovery of G-quadruplex selective ligands" *Methods* 2007, *43*, 313-323.

[56] Mergny, J.L.; Maurizot, J.C. "Fluorescence resonance energy transfer as a probe for G-quartet formation by a telomeric repeat" *Chembiochem* 2001, *2*, 124-132.

[57] De Cian, A.; Guittat, L.; Kaiser, M.; Sacca, B.; Amrane, S.; Bourdoncle, A.; Alberti, P.; Teulade-Fichou, M.P.; Lacroix, L.; Mergny, J.L. "Fluorescence-based melting assays for studying quadruplex ligands" *Methods* 2007, *42*, 183-195.

[58] Rosu, F.; Gabelica, V.; Houssier, C.; De Pauw, E. "Determination of affinity, stoichiometry and sequence selectivity of minor groove binder complexes with double-stranded oligodeoxynucleotides by electrospray ionization mass spectrometry" *Nucleic Acids Res.* 2002, *30*, e82.

[59] Rosu, F.; Gabelica, V.; Shin Ya, K.; De Pauw, E. "Telomestatin-induced stabilization of the human telomeric DNA quadruplex monitored by electrospray mass spectrometry" *Chem. Commun.* 2003, *21*, 2702-2703.

[60] Dixon, I.M.; Lopez, F.; Tejera, A.M.; Estève, J.P.; Blasco, M.A.; Pratviel, G.; Meunier, B. "A G-quadruplex ligand with 10000-fold selectivity over duplex DNA" J. Am. Chem. Soc. 2007, 129, 1502-1503.

[61] White, E.W.; Tanious, F.; Ismail, M.A.; Reszka, A.P.; Neidle, S.; Boykin, D.W.; Wilson, W.D. "Structure-specific recognition of quadruplex DNA by organic cations: influence of shape, substituents and charge" *Biophys. Chem.* 2007, *126*, 140-153.

[62] Zhou, J.; Yuan, G. "Specific recognition of human telomeric G-quadruplex DNA with small molecules and the conformational analysis by ESI mass spectrometry and circular dichroism spectropolarimetry" *Chemistry* 2007, *13*, 5018-5023.

[63] De Cian, A.; Mergny, J.L. "Quadruplex ligands may act as molecular chaperones for tetramolecular quadruplex formation" *Nucleic Acids Res.* 2007, *35*, 2483-2493.

[64] Kern, J.T.; Kerwin, S.M. "The aggregation and G-quadruplex DNA selectivity of charged 3,4,9,10-perylenetetracarboxylic acid diimides" *Bioorg. Med. Chem. Lett.* 2002, *12*, 3395-3398.

[65] Kern, J.T.; Thomas, P.W.; Kerwin, S.M. "The relationship between ligand aggregation and G-quadruplex DNA selectivity in a series of 3,4,9,10-perylenetetracarboxylic acid diimides" *Biochemistry* 2002, *41*, 11379-11389.

[66] Kerwin, S.M.; Chen, G.; Kern, J.T.; Thomas, P.W. "Perylene diimide G-quadruplex DNA binding selectivity is mediated by ligand aggregation" *Bioorg. Med. Chem. Lett.* 2002, *12*, 447-450.

[67] Kim, N.W.; Piatyszek, M.A.; Prowse, K.R.; Harley, C.B.; West, M.D.; Ho, P.L.; Coviello, G.M.; Wright, W.E.; Weinrich, S.L.; Shay, J.W. "Specific association of human telomerase activity with immortal cells and cancer" *Science* 1994, *266*, 2011-2015.

[68] Kim, N.W.; Wu, F. "Advances in quantification and characterization of telomerase activity by the telomeric repeat amplification protocol (TRAP)" *Nucleic Acids Res.* 1997, *25*, 2595-2597.

[69] Gomez, D.; Mergny, J.L.; Riou, J.F. "Detection of telomerase inhibitors based on g-quadruplex ligands by a modified telomeric repeat amplification protocol assay" *Cancer Res.* 2002, *62*, 3365-3368.

[70] Cristofari, G.; Lingner, J. "Telomere length homeostasis requires that telomerase levels are limiting" *EMBO J.* 2006, *25*, 565-574.

[71] De Cian, A.; Lacroix, L.; Douarre, C.; Temime-Smaali, N.; Trentesaux, C.; Riou, J.F.; Mergny, J.L. "Targeting telomeres and telomerase" *Biochimie* 2008, *90*, 131-155.

[72] De Cian, A.; Cristofari, G.; Reichenbach, P.; De Lemos, E.; Monchaud, D.; Teulade-Fichou, M.P.; Shin-Ya, K.; Lacroix, L.; Lingner, J.; Mergny, J.L. "Reevaluation of telomerase inhibition by quadruplex ligands and their mechanisms of action" *Proc. Natl. Acad. Sci USA* 2007, *104*, 17347-17352.

[73] Wilson WD, Sugiyama H. "First International Meeting on Quadruplex DNA" *ACS Chem. Biol.* 2007, *2*, 589-94; Bates P, Mergny JL, Yang D. "Quartets in G-major. The First International Meeting on Quadruplex DNA" *EMBO Rep.* 2007, *8*, 1003-10.

Reviewed by

Prof. Concetta Giancola, University of Naples "Federico II"

Ligands, Polymers and Amino Acids
Editor: Tsisana Shartava

ISBN: 978-1-61122-793-2

EVALUATION OF A SYNTHETIC BIOMIMETIC LIGAND FOR THE PURIFICATION OF THERAPEUTIC PROTEINS

Dimitris Platis and Nikolaos E. Labrou*
Agricultural University of Athens, Athens, Greece,

ABSTRACT

The successful commercialization of high-purity proteins and enzymes parallels the development and improvement of downstream processing technology. Biomimetic ligands find wide application in protein separation and purification, being the most promising affinity ligands of large-scale potential. In the present paper we evaluate the ability of the biomimetic ligand 4-amino-phenyl-oxanilic acid coupled to Sepharose CL-6B via 1,3,5-trichloro-2,4,6-triazine to bind and purify human monoclonal anti-HIV antibody 2F5 (mAb 2F5) from spiked maize extracts and recombinant influenza virus neuraminidase (NA) expressed in insect cells. Under selected conditions the affinity adsorbent exhibited high selectivity and purifying ability for mAb 2F5 and NA.

Keywords: Affinity chromatography; antibody purification; recombinant proteins; therapeutic proteins

ABBREVIATIONS

BEVS, baculovirus expression vector system; mAb 2F5, human anti-HIV 2F5 monoclonal antibody; NA, neuraminidase; SDS-PAGE, sodium dodecyl polyacrylamide gel electrophoresis

1. INTRODUCTION

The use of plants as molecular factories (bioreactors) for the production of therapeutic proteins has a number of advantages, including the lack of pathogenic contaminants, low production cost, and potential for scale-up [1-6]. There are many examples of fully

* To whom all correspondence should be addressed: Laboratory of Enzyme Technology, Department of Agricultural Biotechnology, Agricultural University of Athens, 75 Iera Odos, GR 118 55 Athens, Greece, E-mail: lambrou@aua.gr, Tel./fax +30 210 5294308

functional mammalian proteins expressed in plants, ranging from proinsulin [2] to multimeric antibodies [3].

Baculoviruses are highly efficient eukaryotic expression vectors for producing recombinant proteins in cultured insect cells [7]. Baculovirus-expressed proteins appear to be correctly folded and processed in the majority of cases, even when the protein is fairly large. This is not the case with proteins expressed in prokaryotic and lower eukaryotic systems. Additionally, insect cells are capable of many of the post-translational modifications that occur in mammalian cells, such as glycosylation, phosphorylation, acylation, and amidation [8]. Glycosylation in insect cells appears to employ similar mechanisms as those used in mammalian cells. Although the carbohydrate moieties added to proteins in insect cells appear to be less complex than those on their mammalian cell-expressed counterparts, the immunogenicity of insect cell-expressed and mammalian cell-expressed glycoproteins appear to be equivalent. Finally, baculovirus-expressed proteins usually self-assemble into the higher-order structures normally assumed by the natural proteins.

To take full advantage of the significant advantages of plant and insect cell bioreactors for the production of therapeutic proteins it is necessary that downstream purification of the recombinant product is accomplished economically since downstream processing can account for 50-80 % of the total cost in a large-scale production line. Proteins and other macromolecules of interest can be purified from crude extracts or other complex mixtures by a variety of methods. Affinity chromatography makes use of specific binding interactions between molecules. A particular ligand is chemically immobilized or "coupled" to a solid support so that when a complex mixture is passed over the column, only those molecules having specific binding affinity to the ligand are purified. Application of chromatography methods in downstream processing, particularly when high purity is required, is extremely important. Affinity chromatography is currently the most refined and specialized purification method for biomolecules.

Although affinity chromatography has found successful applications in protein purification, it suffers from problems regarding ligand stability and cost. Some of the most recent advances in this area have explored the power of rational and combinatorial approaches for designing high selective and stable synthetic affinity ligand the so-called biomimetic ligands. In the present report, we describe the design, synthesis and application of an affinity adsorbent with the immobilized biomimetic ligand 4-amino-phenyl-oxanilic acid. The adsorbent was evaluated for its ability to bind and purify human monoclonal antibody 2F5 (mAb 2F5) from spiked maize extract and recombinant influenza virus neuraminidase (NA) expressed in insect cells.

The human monoclonal antibody 2F5 neutralizes a broad array of primary HIV-1 strains [9-11], and its passive infusion, in combination with other neutralizing mAbs, can prevent infection. This antibody represents a promising therapeutic agent, and the development of an efficient downstream processing protocol for mAb 2F5 purification is a major goal of HIV vaccine programs. The immunizing effect of viral NA has been well documented in experimental animals and the importance of the purified recombinant protein has been shown [12-15]. It has been demonstrated that a baculovirus-derived recombinant NA, in the absence of other influenza virus proteins, can induce NA-specific antibodies, reduce the replication of both homologous and heterovariant virus in mice, and suppress disease [16].

Therefore, the development of an efficient purification protocol for recombinant NA is both of academic interest and practical importance.

2. EXPERIMENTAL

2.1. Materials

Epichlorohydrine, 1,6-diamine hexane and 1,3,5-trichloro-2,4,6-triazine, 4-amino-phenyl-oxanilic acid and immuno-chemicals were obtained from Sigma-Aldrich. Purified human monoclonal mAb 2F5 was a much appreciated gift from Prof. A. Katinger and Dr. G. Stiegler (Polymun Scientific Inc, Austria). Recombinant neuraminidase (H1N1, A/Beijing/262/95) expressed in *expres*SF+® insect cells was a much appreciated gift from Protein Sciences Corporation (Meiden, USA). All other chemicals were of analytical purity and were obtained from Merck (Darmstadt, Germany).

2.2. Methods

2.2.1 Synthesis of Affinity Adsorbents

4-Amino-phenyl-oxanilic acid immobilization on Sepharose CL 6B was carried out according to the general procedure published by Platis *et al.*, 2006 [17] as follows:

Activation of agarose. Sepharose CL-6B beads (10 g) were thoroughly washed with double distilled water in a glass filter funnel and drained. The washed beads were suspended in 1M NaOH solution (20 mL) and shaken on a rotary shaker (140 rpm, 30 °C, 10 min). Subsequently, 10 mL epichlorohydrin solution (1 M) were added and the mixture was shaken for an additional 2 h (140 rpm, 30 °C). The beads were thoroughly washed with double distilled water to remove excess epichlorohydrin.

Determination of epoxy-group density. The epoxy density was determined by titrating the pH from alkali to pH 7.0 with 0.1 M HCl after the addition of sodium thiosulfate (1.3 M, 9 mL) to 3 g of moist weight gel according to Sundberg *et al.* [18]. The amount of epoxy groups present is calculated from the amount of hydrochloric acid required to reach pH 7.0.

Coupling with 1,6-diaminehexane. 1,6-diaminehexane (0.5 g) dissolved in double distilled H_2O (10 mL) was added to epoxy-activated Sepharose. The mixture was shaken at 140 rpm, 30° C for 18 h. After completion of the reaction the 1,6-diaminehexane-derivatized agarose was washed with double distilled water. Ninhydrin test was used to determine the amine ligand density of the 1,6-diaminehexane-derivatized agarose beads.

Activation with 1,3,5-trichloro-2,4,6-triazine. 1,6-diaminehexane-derivatized Sepharose was suspended in acetone/water solution (1:1, 31 mL, 0° C) and maintained at 0 ° C in an ice-NaCl bath. 1,3,5-trichloro-2,4,6-triazine (200 mg) dissolved in acetone (100%, 5 mL, 0° C) was added portionwise every 10 min over a period of 2 h with shaking. After completion of the reaction, the gel was washed sequentially with 50 mL of acetone-distilled water mixtures [1:3, 1:1, 1:0, 3:1, 1:1, 1:3 and 0:3 (v/v)] to remove unreacted 1,3,5-trichloro-2,4,6-triazine.

Coupling with 4-amino-phenyl-oxanilic acid. 5 g of dichlorotriazinyl-activated Sepharose was resuspended in a phosphate buffer solution (0.1 M, pH 8, 20 mL) along with 4-amino-phenyl-oxanilic acid (0.4 mmol dissolved in 10 mL water; the pH was adjusted to 8 with 0.1 NaOH). The mixture was shaken on a rotary shaker (210 rpm, 30 °C) for 4 h. After completion of the reaction, the gel was heated at 70 °C to hydrolyze the free chlorine on the triazine ring. Finally the gel filtered, washed with distilled water and stored in water at 4 °C.

2.2.2. Assay of Neuraminidase Activity.

Enzyme assays for fetuin hydrolysis were performed at 30 °C according to the published methods [19], using a Hitachi U-2000 double beam UV-Vis spectrophotometer equipped with a thermostated cell-holder (10 mm path length). One unit of enzyme activity is defined as the amount that catalyses the liberation of 1 μmol of α-sialic acid per minute. Observed reaction velocities were corrected for spontaneous reaction rates when necessary.

2.2.3. Protein Extraction

Protein extraction from maize was carried out according to Platis *et al.*, 2008 [20]. Protein extraction from *expres*SF+® insect cells was carried out according to Dalakouras *et al.*, 2006 [21].

2.2.4. Absorption Equilibrium Studies

In a total volume of 0.5 mL of 20 mM Tris-HCl, pH 7.2, varying amounts of pure mAb (2.5-320 μg/mL), previously dissolved in 20 mM Tris-HCl, pH 7.2, were mixed with 10 mg of affinity adsorbent. The suspensions were shaken for 120 min in order for the system to reach equilibrium. The mixture was then centrifuged (5,000 g, 2 min) and the amount of unbound proteins in the supernatant was determined by Bradford assay [22]. Bound protein was calculated by subtracting the amount of unbound protein from the total amount of protein added.

2.2.5. Purification of Mab 2F5 from Spiked Maize Extract

Spiked maize extract [maize seed extract, 50 μg total protein/mL, 5 % w/w mAb 2F5, in 10 mL of 20 mM Tris-HCl, pH 7.5] was loaded on the immobilized 4-amino-phenyl-oxanilic adsorbent (0.5 mL moist adsorbent). The column was washed with equilibration buffer prior to elution with equilibration buffer containing increasing concentrations of KCl (20-100 mM) in equilibration buffer. The flowthrough and eluted fractions were collected, analyzed for the presence of mAb 2F5 by ELISA (according to Platis *et al.* [17,23]) and total protein was determined by Bradford assay [22].

2.2.6. Purification of Recombinant Influenza Virus Neuraminidase Expressed in Expressf+® Insect Cells.

Dialyzed extract (0.5 mL) was applied to a column prepacked with immobilized biomimetic ligand, previously equilibrated with 5 mL dialysis buffer (25 mM Tris-HCl, pH 7, 2 mM $CaCl_2$, 0.5% (v/v) Triton X-100). The column was initially washed with dialysis buffer (5 mL) and then with 0.2M KCl (5 mL). Finally, elution was accomplished with 50% (v/v) glycerol (5 mL). The flowthrough, wash and elution fractions were collected and the enzymatic activity and protein concentration were measured.

2.2.7. Sodium Dodecyl Sulphate Polyacrylamide Gel Electrophoresis (SDS-PAGE)

SDS-PAGE was carried out as described in [17] and [24]. Protein purity in SDS-PAGE gels was determined by densitometry using the 1D Image Analysis Software (version 3.5; Kodak).

3. RESULTS

3.1. Design and Synthesis of Affinity Adsorbent.

Aromatic amino acid derivatives have been used as bioaffinity ligands in affinity chromatography. They are small and simple molecules with high chemical stability and low cost. Mixed-type biomimetic affinity ligands, which are ligands that provide a combination of both hydrophobic and polar functions, have been previously shown to perform successfully in the isolation of many therapeutic proteins and especially antibodies. It is well established that hydrophobic and electrostatic interactions are the major driving force for mAbs adsorption on several affinity adsorbents [17,25].

Before ligand design, biocomputing analysis was carried out aiming at analyzing the presence of a possible binding pocket on mAb2F5, able to accommodate large affinity ligands. Analysis of the mAb 2F5 structural model with CASTp [26] revealed the presence of several large pockets along with numerous smaller ones. The larger pockets, which are more likely to participate in protein-ligand interactions, with a solvent accessible surface area ranging from 103.9 to 728.8 $Å^2$ (Richards' surface) are shown in Figure 1B. Hydrophobicity [27] and Coulomb electrostatic potential analysis [28] revealed that these pockets provide a mixed-type environment with high hydrophobicity and positive electrostatic potential, thus providing favorable conditions of accommodating negatively charged hydrophobic ligands. This led us to evaluate a library of biomimetic affinity ligands for their ability to bind mAb 2F5. Following affinity screening of the library (data not shown) one biomimetic ligand, bearing 4-amino-phenyl-oxanilic acid coupled to 1,3,5-trichloro-2,4,6-triazine (Figure 1A), was selected for further study.

The synthesis of the affinity adsorbent was implemented directly on the solid phase [29,30] as shown in Figure 2. The selection of 1,3,5-triazine scaffold was based on its synthetic accessibility. In general, the three chlorine atoms of 1,3,5-triazine can be displaced in a sequential and controlled fashion in aqueous solution at approximately 0, 30–50 and 60–100 °C, respectively. Other advantages of 1,3,5-triazine scaffold are its low cost and high stability in neutral buffer solutions. The presence of electron rich nitrogen sites in triazine ring increases its capability of forming additional interactions (e.g. hydrogen bonds, amino-aromatic interactions) with amino acid residues within protein binding site [31]. The first highly reactive chlorine on trichlorotriazine was displaced at 0 °C (2 h) with amino-derivatized Sepharose 6B to yield dichlorotriazinyl-activated Sepharose 6B, whence sequential substitution of the second chlorine on the dichlorotriazinyl-activated Sepharose with 4-amino-phenyl-oxanilic acid and the third chlorine with –OH (by hydrolysis reaction) yielded the affinity adsorbent shown in Figure 3.

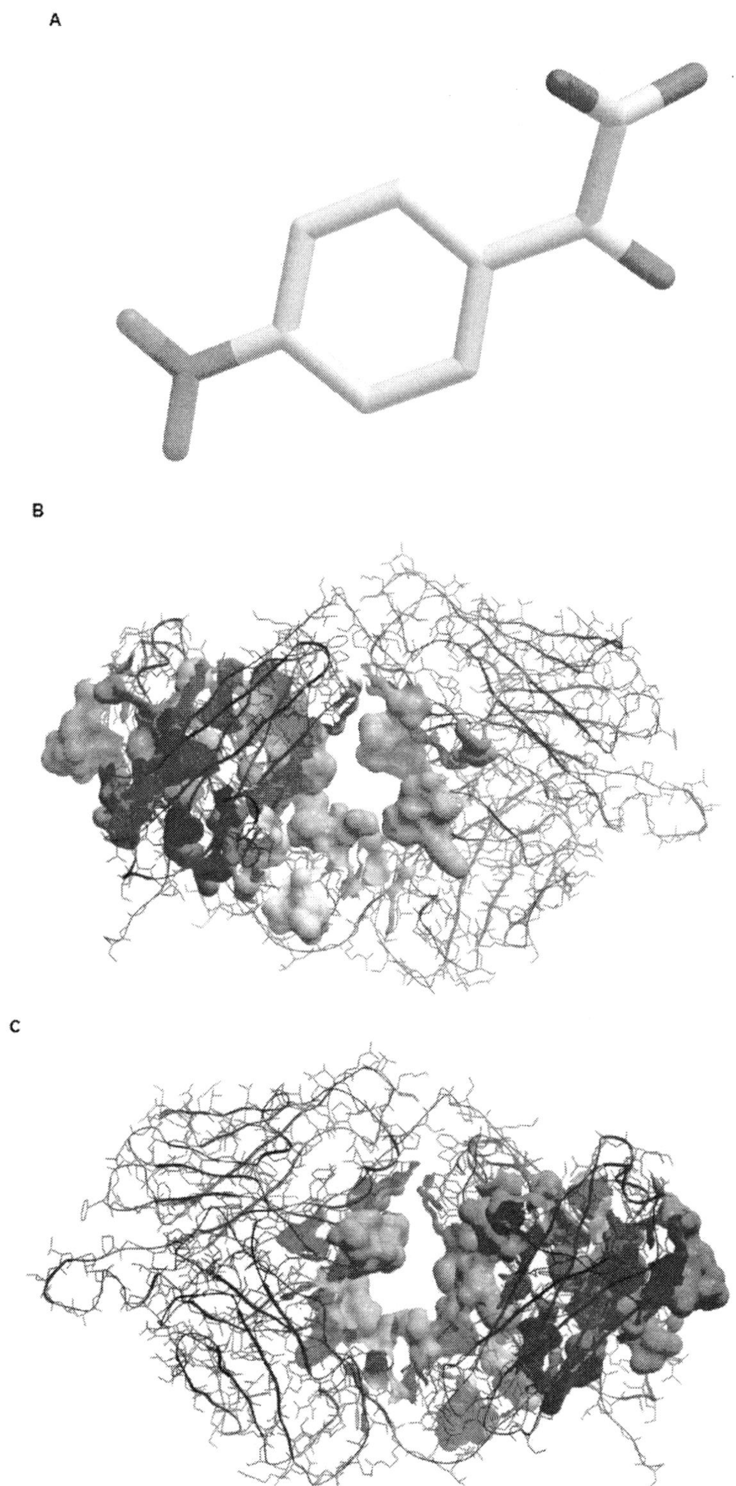

Figure 1. A: The structure of the affinity ligand 4-amino-phenyl-oxanilic acid. B and C: Analysis of the mAb 2F5 structural model with CASTp. The pockets are shown in different colours (B: front view, C: back view).

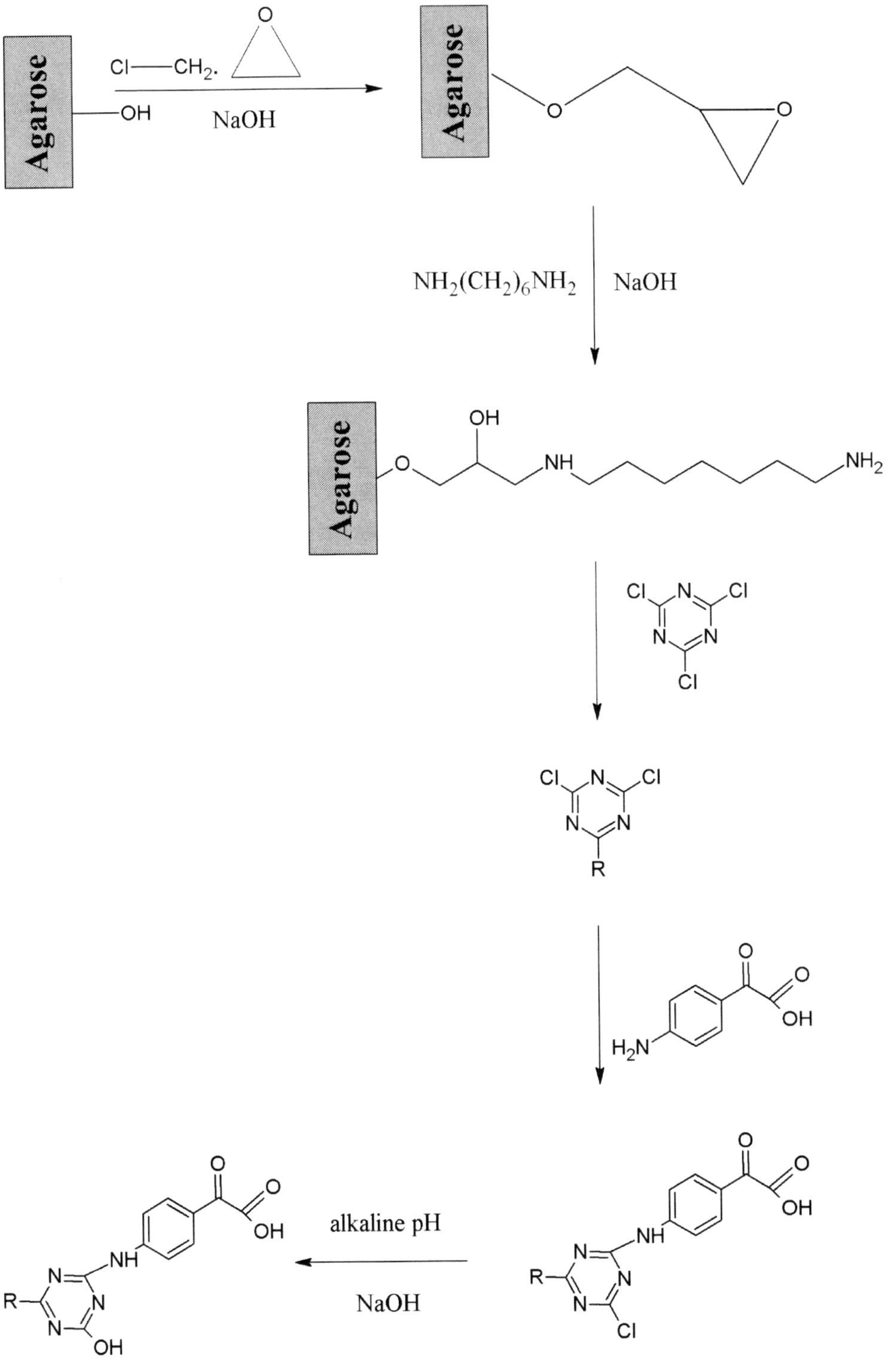
Agarose
Cl—CH2.
OH
NaOH
NH2(CH2)6NH2
NaOH
OH
NH
NH2
Cl
N
R
H2N
OH
alkaline pH
NaOH

Figure 2. The synthetic route for solid phase synthesis of the affinity adsorbent. Where: R is the structure of aminated agarose.

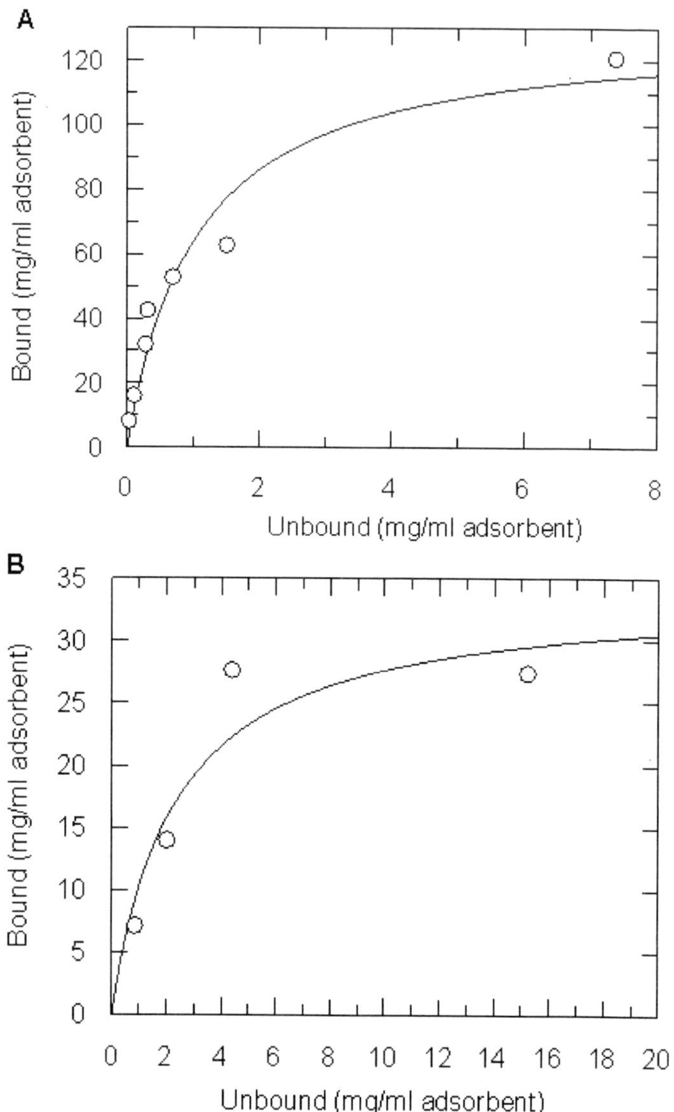

Figure 3. Equilibrium adsorption of mAb 2F5 (A) and IgG fraction (B) on 4-amino-phenyl-oxanilic acid affinity adsorbent. The plots describing the equilibrium in liquid phase protein concentration [Unbound (C*), μg/mL] versus the equilibrium in solid phase protein concentration [Bound (Q*), μg/mL adsorbent].

3.2. Adsorption Equilibrium Studies

Adsorption equilibrium studies were used to characterize the interaction of the mAb 2F5 with the immobilized adsorbent. This approach provides a relationship between the concentration of the protein in the solution and the amount of protein adsorbed to the solid phase, when the two phases are at equilibrium [21,32]. The model most often employed for affinity-ligand systems is the second-order reversible interaction, where the enzyme is assumed to interact with the ligand by a monovalent interaction (equation 1) that has a characteristic binding energy:

$$P + L \underset{k_2}{\overset{k_1}{\rightleftharpoons}} P\text{-}L \qquad \text{equation (1)}$$

where P is the protein in solution, L is the ligand, and P-L is the protein-ligand complex. The parameters k_1 and k_2 are the forward and reverse rate constant, respectively, for the adsorption process. The mass balance equation formulation requires an independent description of the adsorption kinetics. This is commonly obtained by fitting an adsorption isotherm equation to measured data. Although numerous expressions exist, the Langmuir isotherm (equation 2) is by far the most applied [17,21,32]:

$$\frac{dQ}{dt} = k_1 C(Q_{\max} - Q) - k_2 Q \text{ equation} \qquad (2)$$

where, Q is the amount of target protein adsorbed, Q_{max} the maximum adsorption capacity. At equilibrium, the following relationship is obtained:

$$Q^* = \frac{Q_{\max} C^*}{K_D + C^*} \text{ equation} \qquad (3)$$

where Q^* and C^* are the equilibrium concentrations of the adsorbed and bulk-phase target molecules respectively, and K_D is the equilibrium dissociation constant, i.e. k_2/k_1.

The equilibrium adsorption data, for the batch adsorption of mAb 2F5 on the affinity adsorbent, were well fitted by a Langmuir isotherm. A typical fit is shown in Figure 3A. The dissociation constant (K_D, Table 1) was determined to be 0.12 μM, within the range would be expected for a high-affinity ligand [17,21,32]. The theoretical binding capacity of the adsorbent was 130.8 mg/mL moist adsorbent. The Langmuir adsorption model assumes that the molecules are adsorbed at a fixed number of well-defined sites, each of which can only hold one molecule. These sites are also assumed to be energetically equivalent, and distant to each other so that there are no interactions between molecules adsorbed to adjacent sites [32]. This isotherm was originally developed for the next-neighbour independent monolayer adsorption of ideal gases [33,34]. Application of the model to protein solutes in a solvent makes the associated parameters dependent upon the state of the solvent. Although the assumptions behind the Langmuir isotherm are not strictly valid for proteins the Langmuir equation has been applied, nevertheless, with reasonable success in many cases [17,20,35]. Equilibrium adsorption data, for the batch adsorption of polyclonal antibody fraction (Sigma-Aldrich) on the affinity adsorbent, were also well fitted by a Langmuir isotherm (Figure 3B, Table 1). This indicates that the affinity adsorbent might be of general use for the purification of antibodies of different specificities.

Table 1. Dissociation constants and theoretical binding capacities of mAb 2F5 and IgG fraction for the affinity adsorbents as determined by absorption equilibrium studies.

Biomimetic affinity adsorbent	mAb 2F5	IgG fraction
Dissociation constant, K_D (μM)	0.12±0.03	0.32±0.04
Theoretical capacity, Q_{max} (mg/mL moist adsorbent)	130.8	37.2

3.3. Purification of mAB 2F5 from Spiked Maize Extract

Spiked maize extract (maize seed extract at 50 μg total protein/mL with 5 % w/w mAb 2F5 in 10 mL, 20 mM Tris-HCl buffer, pH 7.5) was applied directly to the adsorbent. Unbound proteins were removed with equilibration buffer prior to elution of mAb 2F5 with equilibration buffer containing increasing concentrations of KCl (20-100 mM) with a recovery of > 90 %.

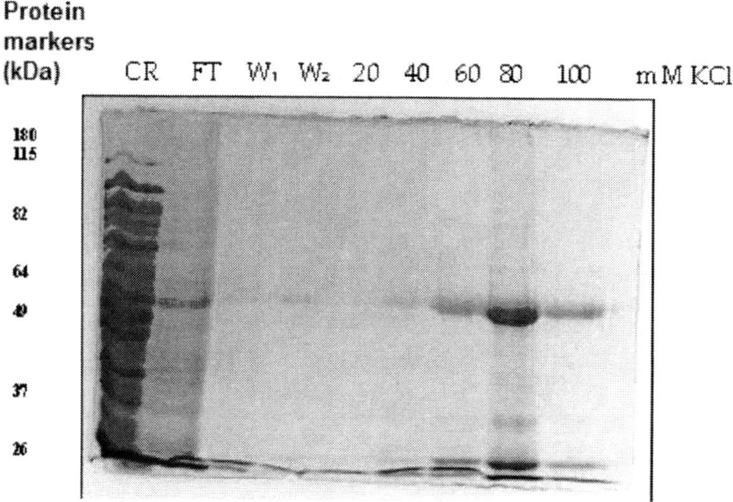

Figure 4. A. SDS-Polyacrylamide gel electrophoresis of mAb 2F5 antibody purification from crude corn extract spiked with mAb 2F5. The protein bands were stained with Coomassie Blue R-250. CR: crude spiked corn extract; FT: flowthrough fraction; W1 and W2: wash fractions; 20-100 mM KCl: eluted fractions from the affinity column with different concentrations of KCl.

SDS-PAGE analysis of the eluted antibody fraction was carried out and the results are illustrated in Figure 4. Analysis of the purity of mAb 2F5 by reducing SDS-PAGE showed that the eluted fraction had a purity > 90 %. Purity was determined by densitometry using the 1D Image Analysis Software (version 3.5; Kodak). In addition, mAb 2F5 was not subject to degradation during the course of the purification protocol, a problem often encountered during purification of antibodies. The procedure is simple and effective yielding monoclonal antibody 2F5 of high purity and good recovery. An additional polishing step, such gel-filtration, should render the final product clinical grade and ready for human testing. It is noteworthy that the recovery of 2F5 monoclonal antibody throughout the entire chromatographic procedure is accomplished in mild, non-denaturing chromatographic conditions, which preserve the native structure and protect the functionality of the antibody, a significant attribute for therapeutic applications.

3.4. Application of the Affinity Adsorbent to the Purification of Influenza Virus Neuraminidase

To further demonstrate the effectiveness of the biomimetic adsorbent, the chromatographic behaviour of influenza virus neuraminidase (NA) was evaluated. NA (H1N1, A/Beijing/262/95) was expressed in in *expres*SF+® insect cells using the baculovirus expression vector system (BEVS) [21]. BEVS provides an excellent method for the development of the ideal subunit vaccine for a variety of reasons. Baculovirus expression of recombinant proteins is produced in approximately eight weeks. This is especially critical in addressing pandemic threats. Baculoviruses are safe by virtue of their narrow host range, which is restricted to a few taxonomically related insect species. They have not been observed to replicate in mammalian cells [36, 37]. Finally, because the insects infected by baculoviruses are non-biting, humans generally do not have pre-existing immunity to insect cell proteins which could cause an allergic reaction to trace amounts of insect cell proteins in the vaccine preparation [38].

The immobilized biomimetic ligand 4-amino-phenyl-oxanilic acid binds reversibly to influenza virus neuraminidase with a capacity of 5.3 mU/mL adsorbent and being eluted using a solution of 50% (v/v) glycerol with approximately 50 % recovery and purity 70-80%. These results suggest that the ligand may be of wider use and can be exploited for the purification of different proteins.

CONCLUSION

A major challenge in the biotechnology industry is the production and purification of biologically active recombinant therapeutic proteins. Cell-based therapeutic proteins and vaccines have been used for more than 20 years to prevent or cure deceases since they are reliable and effective. However, their drawbacks are high cost and lengthy production cycle. The mAb 2F5 and NA are promising therapeutic candidates. However, their usefulness ultimately depends on the ability to produce enough of these proteins to satisfy demand. The work reported in the present report demonstrates a progress toward the development of effective, reliable and economical downstream processing protocol for recombinant protein-based vaccines. The methods hold potential for application at preparative scale, since it is simple and employs low-cost materials. Affinity chromatography using biomimetic adsorbents has shown to be extremely valuable for purification of biopharmaceuticals and will remain the preferred technique for the future.

ACKNOWLEDGMENTS

The authors thank the European Union for the financial assistance provided. This work was performed within the framework of the Pharma-Planta research contract (No 503565).

REFERENCES

[1] Evangelista,R.L.; Kusnadi,A.R.; Howard,J.A.; Nikolov,Z.L. Process and economic evaluation of the extraction and purification of recombinant beta-glucuronidase from transgenic corn. *Biotechnol.Prog.*, 1998, 14, 607-614

[2] Farinas,C.S.; Leite,A.; Miranda,E.A. Aqueous extraction of recombinant human proinsulin from transgenic maize endosperm . *Biotechnol.Prog.*, 2005, 21, 1466-1471

[3] Ma,J.K.; Drake,P.M.; Christou,P. The production of recombinant pharmaceutical proteins in plants . *Nat.Rev.Genet.*, 2003, 4, 794-805

[4] Stoger,E.; Sack,M.; Nicholson,L.; Fischer,R.; Christou,P. Recent progress in plantibody technology . *Curr.Pharm.Des*, 2005, 11, 2439-2457

[5] Stoger,E.; Ma,J.K.; Fischer,R.; Christou,P. Sowing the seeds of success: pharmaceutical proteins from plants . *Curr.Opin.Biotechnol.*, 2005, 16, 167-173

[6] Fischer,R.; Stoger,E.; Schillberg,S.; Christou,P.; Twyman,R.M. Plant-based production of biopharmaceuticals . *Curr.Opin.Plant Biol.*, 2004, 7, 152-158

[7] Ikonomou L, Schneider YJ, Agathos SN. Insect cell culture for industrial production of recombinant proteins. *Appl Microbiol Biotechnol.* 2003 Jul;62(1):1-20.

[8] Altmann F, Staudacher E, Wilson IB, Marz L. Insect cells as hosts for the expression of recombinant glycoproteins. *Glycoconj* J. 1999 Feb;16(2):109-23.

[9] Conley,A.J.; Kessler,J.A.; Boots,L.J.; Tung,J.S.; Arnold,B.A.; Keller,P.M.; Shaw,A.R.; Emini,E.A. Neutralization of divergent human immunodeficiency virus type 1 variants and primary isolates by IAM-41-2F5, an anti-gp41 human monoclonal antibody . *Proc.Natl.Acad.Sci.U.S.A*, 1994, 91, 3348-3352

[10] D'Souza,M.P.; Livnat,D.; Bradac,J.A.; Bridges,S.H. Evaluation of monoclonal antibodies to human immunodeficiency virus type 1 primary isolates by neutralization assays: performance criteria for selecting candidate antibodies for clinical trials. AIDS Clinical Trials Group Antibody Selection Working Group . *J.Infect.Dis.*, 1997, 175, 1056-1062

[11] Trkola,A.; Pomales,A.B.; Yuan,H.; Korber,B.; Maddon,P.J.; Allaway,G.P.; Katinger,H.; Barbas,C.F., III; Burton,D.R.; Ho,D.D.; . Cross-clade neutralization of primary isolates of human immunodeficiency virus type 1 by human monoclonal antibodies and tetrameric CD4-IgG . *J.Virol.*, 1995, 69, 6609-6617

[12] Brett,I.C. and Johansson,B.E. Immunization against influenza A virus: Comparison of conventional inactivated, live-attenuated and recombinant baculovirus produced purified hemagglutinin and neuraminidase vaccines in a murine model system . *Virology*, 2005, 339, 273-280

[13] Deroo,T.; Jou,W.M.; Fiers,W. Recombinant neuraminidase vaccine protects against lethal influenza . *Vaccine*, 1996, 14, 561-569

[14] Johansson,B.E.; Price,P.M.; Kilbourne,E.D. Immunogenicity of influenza A virus N2 neuraminidase produced in insect larvae by baculovirus recombinants. *Vaccine*, 1995, 13, 841-845

[15] Johansson,B.E. Immunization with influenza A virus hemagglutinin and neuraminidase produced in recombinant baculovirus results in a balanced and broadened immune response superior to conventional vaccine. *Vaccine*, 1999, 17, 2073-2080

[16] Hood,E.E. From green plants to industrial enzymes. *Enzyme and Microbial Technol.*, 2002, 30, 279-283

[17] Platis,D.; Sotriffer,C.A.; Clonis,Y.; Labrou,N.E. Lock-and-key motif as a concept for designing affinity adsorbents for protein purification. *J Chromatogr.A*, 2006, 1128, 138-151

[18] Sundberg,L. and Porath,J. Preparation of adsorbents for biospecific affinity chromatography. Attachment of group-containing ligands to insoluble polymers by means of bifuctional oxiranes . *J.Chromatogr.*, 1974, 90, 87-98

[19] Warren,L. The thiobarbituric acid assay of sialic acids. *J.Biol.Chem.*, 1959, 234, 1971-1975

[20] Platis,D. and Labrou,N.E. Affinity chromatography for the purification of therapeutic proteins from transgenic maize using immobilized histamine . *J Sep.Sci.*, 2008, 31, 636-645

[21] Dalakouras,T.; Smith,B.J.; Platis,D.; Cox,M.M.; Labrou,N.E. Development of recombinant protein-based influenza vaccine. Expression and affinity purification of H1N1 influenza virus neuraminidase . *J Chromatogr.A*, 2006, 1136, 48-56

[22] Bradford,M.M. A rapid and sensitive method for the quantitation of microgram quantities of protein utilizing the principle of protein-dye binding. *Anal.Biochem.*, 1976, 72, 248-254

[23] Platis,D. and Labrou,N.E. Development of an aqueous two-phase partitioning system for fractionating therapeutic proteins from tobacco extract. *J.Chromatogr.A*, 2006, 1128, 114-124

[24] Laemmli,U.K. Cleavage of structural proteins during the assembly of the head of bacteriophage T4 . *Nature*, 1970, 227, 680-685

[25] Li,R.; Dowd,V.; Stewart,D.J.; Burton,S.J.; Lowe,C.R. Design, synthesis, and application of a protein A mimetic . *Nat.Biotechnol.*, 1998, 16, 190-195

[26] Liang,J.; Edelsbrunner,H.; Woodward,C. Anatomy of protein pockets and cavities: measurement of binding site geometry and implications for ligand design . *Protein Sci.*, 1998, 7, 1884-1897

[27] Perlman,D. and Halvorson,H.O. A putative signal peptidase recognition site and sequence in eukaryotic and prokaryotic signal peptides . *J Mol.Biol.*, 1983, 167, 391-409

[28] Guex,N. and Peitsch,M.C. SWISS-MODEL and the Swiss-PdbViewer: an environment for comparative protein modeling . *Electrophoresis*, 1997, 18, 2714-2723

[29] Li,R.; Dowd,V.; Stewart,D.J.; Burton,S.J.; Lowe,C.R. Design, synthesis, and application of a protein A mimetic . *Nat.Biotechnol.*, 1998, 16, 190-195

[30] Teng,S.F.; Sproule,K.; Husain,A.; Lowe,C.R. Affinity chromatography on immobilized "biomimetic" ligands. Synthesis, immobilization and chromatographic assessment of an immunoglobulin G-binding ligand . *J Chromatogr.B Biomed.Sci.Appl.*, 2000, 740, 1-15

[31] Labrou,N.E. Design and selection of ligands for affinity chromatography . *J.Chromatogr.B Analyt.Technol.Biomed.Life Sci.*, 2003, 790, 67-78

[32] Livingston,A.G. and Chase,H.A. Preparation and characterization of adsorbents for use in high-performance liquid affinity chromatography. *J Chromatogr.A*, 1989, 481, 159-174

[33] Langmuir,I. The adsorption of gases on plane surfaces of glass, mica and platinum . *J.Am.Chem.Soc.*, 1918, 40, 1361-1403

[34] Langmuir,I. Vapor pressures, evaporation, condensation and adsorption . *J.Am.Chem.Soc.*, 1932, 54, 2798-2832

[35] Karlsson,D.; Jakobsson,N.; Axelsson,A.; Nilsson,B. Model-based optimization of a preparative ion-exchange step for antibody purification. *J Chromatogr.A*, 2004, 1055, 29-39

[36] Hartig PC, Chapman MA, Hatch GG, Kawanishi CY. Insect virus: assays for toxic effects and transformation potential in mammalian cells. *Appl Environ Microbiol.* 1989 55(8):1916-20.

[37] Hartig PC, Cardon MC, Kawanishi CY. Generation of recombinant baculovirus via liposome-mediated transfection. *Biotechniques*. 1991 Sep;11(3):310, 312-3.

[38] Brett,I.C. and Johansson,B.E. Immunization against influenza A virus: comparison of conventional inactivated, live-attenuated and recombinant baculovirus produced purified hemagglutinin and neuraminidase vaccines in a murine model system . *Virology*, 2005, 339, 273-280

Ligands, Polymers and Amino Acids
Editor: Tsisana Shartava
ISBN: 978-1-61122-793-2

PHYTOESTROGENS' BIOCHEMICAL ASPECTS AND BIOLOGICAL ACTIVITIES

Rita de Cássia da Silveira e Sá and Luciana Valente Borges
Federal University of Juiz de Fora, Juiz de Fora, MG, Brazil

ABSTRACT

Phytoestrogens are natural compounds found in plants that possess estrogen-like activity. The major classes of phytoestrogens are isoflavones, lignans and coumestans. Isoflavones are structurally similar to mammalian estrogen and are found in large amounts in soybean and soy products. Lignans are found in flaxseed, whole grain cereals, teas and some vegetables, and coumestans are mainly present in clover and alfalfa sprouts. The major isoflavones are genistein, daidzein and glycitein. These compounds are hydrolysis products of glycosides in plants, termed genistin, daidzin and glycitin, which, after oral ingestion, are metabolized by intestinal flora. In lignans, the plant precursors are matairesinol and secoisolariciresinol, which are also metabolized by the intestinal microflora to enterolactone and enterodiol, respectively. Phytoestrogens have been largely used in the world and for a long time period, and it has been observed that the incidence of hormone-dependent diseases is reduced in countries, especially Asian countries, with a high dietary content of phytoestrogens. Animal models and clinical investigations have shown that phytoestrogens are protective in the prevention of hormone-dependent cancers, primarily breast and prostate cancers; cardiovascular disease; osteoporosis and to alleviate the symptoms of menopause. The estrogenic effects of the phytoestrogens are mediated through the estrogen receptors ERα and ERβ, which function as ligand-inducible transcription factors for genes involved in cell growth, proliferation, and differentiation. Many mechanisms of action have been identified for phytoestrogen prevention of diseases, including estrogenic/antiestrogenic activity, antiproliferation, induction of cell cycle arrest and apoptosis, antioxidation activity, induction of detoxification enzymes, regulation of the host immune system, and changes in cell signaling. Despite growing evidence of the health benefits of phytoestrogens, it is important to investigate further their mechanism of actions for a better understanding of the biological outcomes and possible adverse effects.

INTRODUCTION

Numerous epidemiological studies have shown possible associations between diet and many common diseases. It is now generally accepted that many of these diseases, more prevalently found in Western cultures, are diet related [1] and could be avoided with a thorough dietary intervention. Populations and case-control studies have provided a large body of information that correlates diets low in fat and rich in complex carbohydrates from vegetables, fruits and grains to a decreased risk of chronic diseases [1,2,3,4,]. In addition to the essential nutrients, the human diet contains naturally occurring plant-derived bioactive constituents known as phytochemicals. Among these biologically active substances, there are carotenoids, sulphides, coumarins, terpens, saponins, glucosinolates, flavonoids, protease inhibitors, phenolic acids and plant sterols, that, if consumed either naturally as an integral part of the food or as a food supplement, may provide long-term health benefits, such as prevention of degenerative diseases [1,5].

One class of phytochemicals, named phytoestrogens, comprises a group of natural compounds found in plants (e.g. vegetables, legumes, fruits, whole grains, and, specially, soy food products) to varying degrees. Phytoestrogens are classically defined as non-steroidal polyphenolic compounds that have similar chemical and structural properties to that of estrogens [5,6,7,8]. Hundreds of molecules fall under this classification and they can be divided into two main groups: the flavonoids that are further subdivided into isoflavones, coumestans and prenyl flavonoids; and non-flavonoids, comprising the lignans and stillbenes [9] (Figure 1)

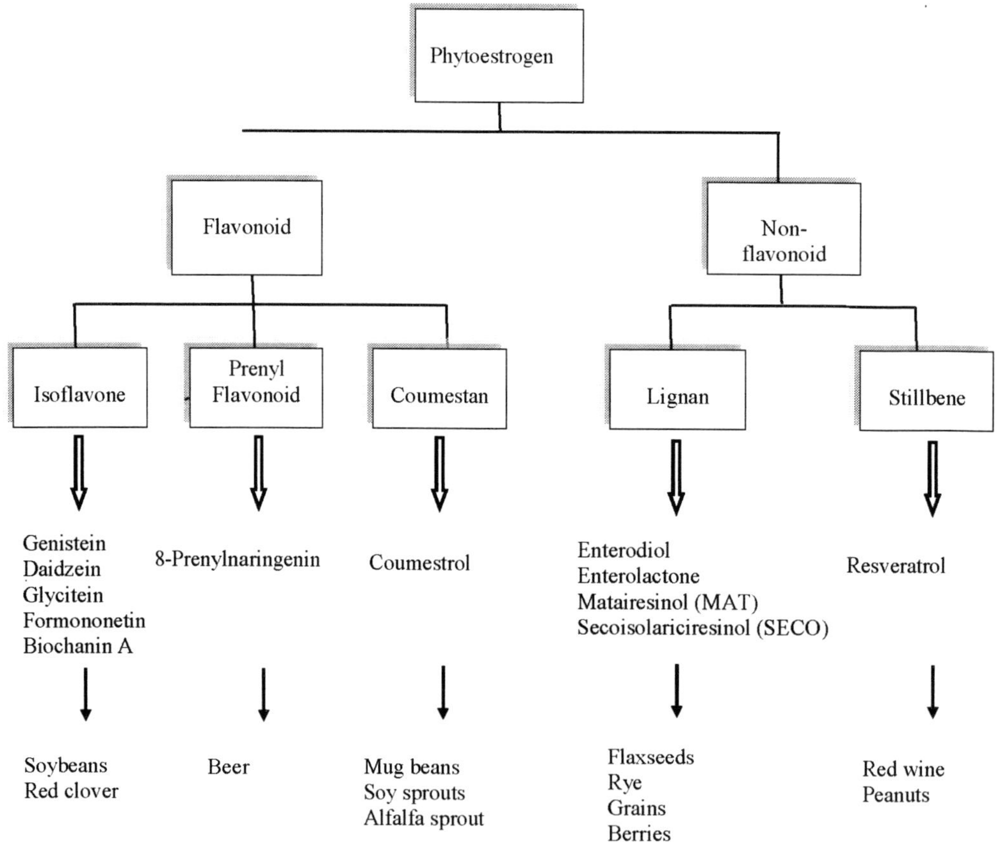

Figure 1. Schematic presentation of the different classes of phytoestrogens. (Based on references 9 and 18.)

The three main classes of phytoestrogens of clinical interest are isoflavones (derived mainly from soy beans and clover), lignans (found in flaxseed in large quantities) and coumestans (derived from sprouting plants like alfalfa) [10]. Of these classes, isoflavones play a major role in human consumption, being found in a variety in foodstuffs, and because of the use of nutritional supplements rich in soy. A fourth class of compounds referred to as resorcylic acid lactones are better classified as myco-estrogens, which are estrogenic fungal products that are not plant derived but are found in pasture grasses and legumes infected by the fungal genus *Fusarium* [11,12].

Investigations developed in animal models and clinical trials have shown that phytoestrogens can function as protective agents against age-related diseases (cardiovascular and osteoporosis), and hormone-dependent cancers (e.g. breast and prostate cancers). They also alleviate the symptoms of menopause, such as hot flushes [5,6,7,8,13,14,15,1,6]. The cancer chemopreventive effects of phytoestrogens are apparently related to several possible mechanisms including their ability to inhibit growth factors, DNA topoisomerase, steroidogenic, and enzymes tyrosine kinase(s), which inhibits cell growth due to non-phosphorylation of proteins required for cell division. Other mechanisms are related to their ability to act as antioxidant and antiangiogenic agents [5,17].

Phytoestrogens bind to estrogen receptors and within each class of compounds, they affect the estrogen-mediated response in different ways. For instance, they may act as estrogen agonists at some tissue-specific targets, whereas in others, they may act as antagonistic agents [16,18,19].

SIGNAL TRANSDUCTION BY ESTROGENS

General Aspects

Estrogens are hydrophobic steroid hormones produced in the ovaries and testis that are involved in the development, differentiation, growth, normal functioning, and malignancies of the organs and tissues of the mammalian reproductive systems, such as the mammary gland, ovary, uterus, vagina, testis, epididymis, and prostate [20,21,22]. In addition to these effects, they exert many biological effects in the body beyond the reproductive system, including the cardiovascular system, where estrogens have certain cardioprotective effects, bone maintenance and the brain [23,24]. Estrogen action in the brain affects the activity and connectivity of certain neuronal cells and, consequently, may interfere with important physiological parameters that regulate animal reproduction, including mood and behavior, gonadotropin production and release from the pituitary, and locomotor activity [25]. It is also believed that estrogens may interfere with nonreproductive events in the brain, such as learning and memory [26,27]. In rodents, they are negative regulators of lymphopoiesis, causing transient involution of the thymus during pregnancy and atrophy during estrogen treatment [28]. They are also responsible for promoting the development of certain types of breast cancer and the understanding of the relationship between estrogen dependency and breast tumors is the key for therapeutic intervention with antiestrogens such as tamoxifen [26] and plant-derived compounds such as phytoestrogens [19].

In the hypothalamic-anterior pituitary-gonad axis (Figure 2), the release of the gonadotropin-releasing hormone (GnRH) by the hypothalamus triggers the secretion of the follicle-stimulating hormone (FSH) and luteinizing hormone (LH) in the anterior pituitary to stimulate the granulosa cells in the ovaries to produce estrogen and progesterone. In the testis, FSH and LH stimulate the Sertoli cells or Leydig cells to produce androgen binding protein (ABP) and testosterone, respectively [29]. Following their synthesis and release, estrogens bind primarily to sex hormone-binding globulin (SHBG) in the plasma and, secondarily, to plasma albumin. Owing to their hydrophobic nature, estrogens cross the plasma membrane of the target cells by simple diffusion; however, before doing so, they first dissociate from the plasma-bound proteins. Once inside the cells, estrogens bind with high affinity and specificity to an intranuclear binding protein, termed estrogen receptor, which functions as ligand-inducible transcription factors for genes involved in cell growth, proliferation and differentiation [30,31,32]. The estrogen-receptor interaction results in conformational changes, enabling the estrogen-receptor complex to bind to specific transcriptional enhancer DNA sequences, also known as estrogen-responsive elements (ERE) [32,33,34]. The dominant form of estrogen in the body is 17β estradiol, however, any compound capable of inducing receptor dimerization and subsequent binding to ERE, can be considered an estrogen [35].

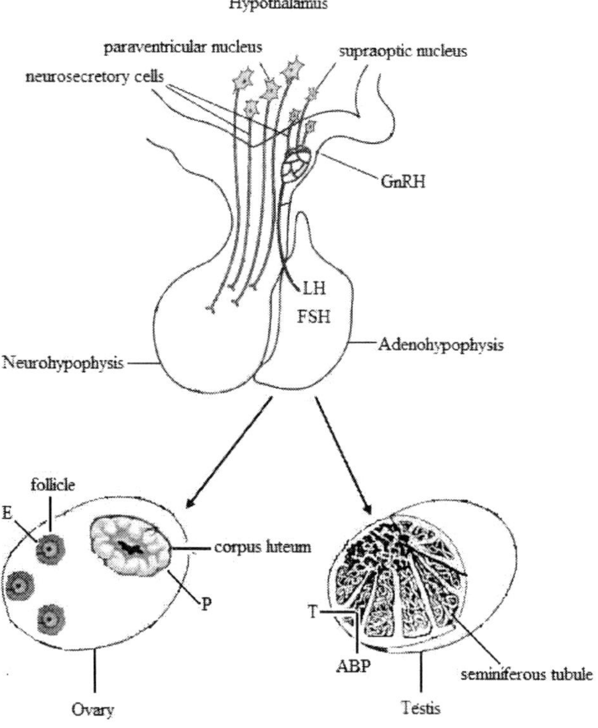

Figure 2. Schematic presentation of the hypothalamus-pituitary-gonads axis. GnRH is released by the hypothalamus to stimulate the adenohypophysis (= anterior hypophysis) to secrete FSH and LH. These hormones stimulate the granulosa cells in the ovaries and the Sertoli cells or Leydig cells in the testis to produce the sex steroid hormones. GnRH = gonadrotopin-releasing hormone; FSH = follicle-stimulating hormone; LH = luteinizing hormone; ABP = androgen binding protein; E = estrogen; P = progesterone; T = testosterone.

The activation process of the estrogen-receptor complex involves the dislocation of heat shock proteins, such as hsp 90, which were blocking the DNA binding site on the receptor [32,34]. Following activation, the estrogen-receptor complex dimer binds to the ERE in the regulating region of the target gene promoters resulting in activation of specific genes, transcription, accumulation of mature mRNAs, and their subsequent translocation into the cytoplasm. In the cytoplasm, mRNAs are translated on the ribosomes and, as a consequence, several specific proteins are synthesized producing various metabolic and physiologic functions [32,33,34] (Figure 3).

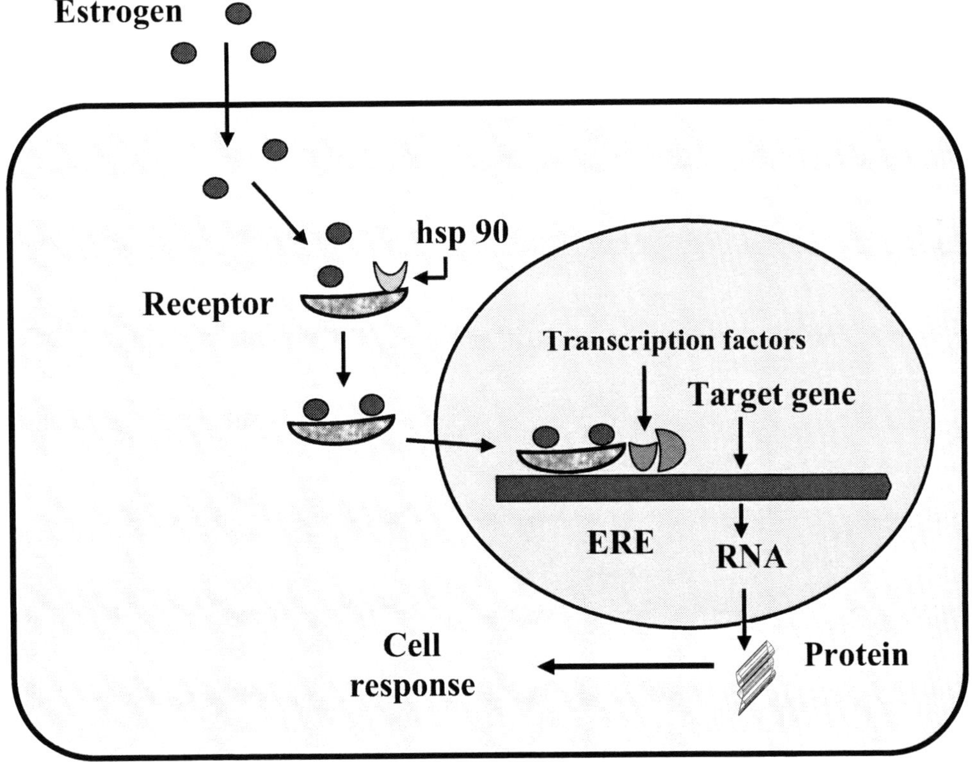

Figure 3. Scheme of a model of estrogen receptor action. Estrogens cross cell membrane by simple or facilitated diffusion and bind to specific receptors. The estrogen receptor (ER) is, in fact, a transcription factor that can be found either in the cytoplasm or the nucleus in an inactive state due to blockage by heat shock proteins (e.g. hsp 90). Upon hsp 90 exits and estrogen binding, ER undergoes activation which enables its effective binding to the estrogen responsive element (ERE) in target gene promoters and activation of transcription of that gene. (Based on reference 24).

The receptor numbers vary greatly in different cell types. For example, reproductive tissues have approximately 100 – 1000 fold greater numbers than bone cells and most of other cells in the body [36]. A small percentage (2-3%) of estrogen receptor are found on the cell membrane [37,38], representing an alternative nongenomic pathway. This pathway has been proposed as an explanation for the fast effects of estrogen in the brain and in the reproductive system [26,39,40,41].

Receptor Structure and Function

The estrogen receptor (ER) is a member of the steroid hormone receptor superfamily of genes which also includes receptors for progestins, glucocorticoids, mineralocorticoids, and androgens [42]. Steroid hormone receptors can be subdivided into several functional domains: 1. a highly variable N-terminal domain (A/B region) that contains a transactivation domain, which interacts with components of the core transcriptional complex; 2. a highly conserved central domain, with a C-region that contains two type II zinc clusters and is responsible for showing specific DNA binding and receptor dimerization [43]; 3. a C-terminal domain (E/F region) which is a functionally complex domain involved in ligand binding and receptor dimerization, nuclear localization, and interactions with transcriptional coactivators and corepressors (44,45).

There are two known estrogen receptor, ERα and ERβ [35]. Although these receptors can be localized within the same cell, they vary in tissue distributions and many have different effects on mixed agonists and antagonists [46]. Upon binding to the receptor, some compounds act as estrogen agonists or antagonists and are referred to as Selective Estrogen Modulators (SERMs) [46]. For example, the antiestrogen tamoxifen is an estrogen antagonist in breast tissue but act as an agonist in the uterus, bone and vascular system [47]. When a compound binds to the receptor in a target cell and antagonistic effects are produced, it is either because dimerization failed to occur or the correct configuration to activate ERE was not attained [35].

The physiological effects of estrogens are mediated predominantly by ERα [48], whereas ERβ may counteract ERα action [49]. However, even though ERβ has lower binding affinity for and activation by endogenous estrogens, some xenoestrogens preferentially bind and activate ERβ [50]. The presence of ERs has been identified in numerous normal and pathological tissues. Both ERα and ERβ are active in normal ovarian follicular development, vascular endothelia cells, myocardial cells, smooth muscle, and breast tissue [46]. ERα acts in bone maturation in males and females, but only ERβ is involved in bone maintenance in females [46]. ERα is more functional in maintaining follicle-stimulating hormone (FSH) and lutenizing hormone (LH) concentrations in blood, whereas ERβ is more active in frontal lobe mediated learning and memory [46]. The ERβ expression was detected in various brain regions (hypothalamus, cerebellum, olfactory lobe, brain stem) and the spinal cord, as well as in the prostate, testis, epididymis, bladder, uterus, lung and pituitary [25].

Gonadal hormones have an important effect on bone maintenance. Ovariectomized adult rats and monkeys showed a decrease in bone ash weight and bone mineral density. In the menopause, estrogen deficiency is the cause of the most rapid phase of bone loss in women. In both cases, the use of exogenous estrogens can prevent these bone changes from happening and can prevent the fractures associated with low bone mass [51]. In men, associations between plasma estrogen levels and bone mineral density have been observed, suggesting that estrogen is involved in maintaining male skeleton as well. Bone loss and fractures associated with low bone mass also occur in men, although at a lower rate than observed in women [51]. The expression of ERα and ERβ have been detected in osteoblastic cells. It was revealed that osteoblastic cells isolated from the bone of neonatal rats expressed a much higher level ERβ mRNA than that of ERα mRNA [52].

Chemically diverse compounds, for instance, estrogens, some androgens, antiestrogens, and xenoestrogens, have estrogenizing activity and interact with the ER. Phytoestrogens also bind to the estrogen receptors; however, the different classes of phytoestrogens show distinct affinity to ERα and ERβ. This differential affinity is probably functionally related to differences in tissue distribution of these two receptor sub-types and possibly to differences in biological activity [50]. Phytoestrogens compete with estrogen for receptor binding at nanomolar (10^{-9} M) concentrations and in higher macromolar (10^{-6} M) concentrations. Depending on the affinity of the phytoestrogen molecule for the estrogen receptor, the binding of a phytoestrogen to the receptor may result in partial activation of the receptor (weak agonistic effect) or the displacement of an estrogen molecule, which may reduce receptor activation (antagonistic or anti-estrogenic effect).

Phytoestrogens: Biology, Structure, Distribution and Metabolism

The phytoestrogens are naturally occurring compounds with a chemical structure possessing a phenolic ring similar to that of estradiol and two hydroxyl groups at positions that allow the correct distance between them to facilitate binding to the estrogen receptor. Because they can act as estrogen agonists or antagonists, phytoestrogens also act like natural SERMs at various tissues throughout the body [50,53,54]. Their ability to behave like estrogen mimics at some tissue-specific targets apparently plays an important part in their health-promoting effects, whereas, in other target tissues, the antagonistic characteristics displayed is comparable to that of tamoxifen or raloxifene, suggesting that SERM activity is sex hormone and gender dependent [55]. Phytoestrogen actions at the cellular and molecular level are influenced by many factors, such as concentration dependency, receptor status, presence or absence of endogenous estrogens, and the type of target organ or cell [1] (Figure 4).

The three major classes of phytoestrogens – isoflavones, lignans and coumestans – are represented by different compounds, which can be found in many foods, particularly leguminous plants, seeds, nuts and berries. Usually, single plants contain more than one type of phytoestrogen, and different plants contain different concentrations of the several phytoestrogen types [18].

In plants, the phytoestrogens function primarily as antioxidants and phytoalexins. Many isoflavones exhibit antimicrobial activity and are believed to play an important role on plant fight microbial disease. Antimicrobial isoflavones can be classified as pre-formed “phytoanticipins” or inducible “phytoalexins” [36]. Different kinds of lignans have been discovered in many plant species and parts, including the roots, leaves, flowers, fruits, seeds and wooden parts. The accumulation of lignans in the core of trees is important for their durability and longevity. Moreover, they also may function as phytoalexins, providing protection for the plant against diseases and pests, and may participate in controlling plant growth [56].

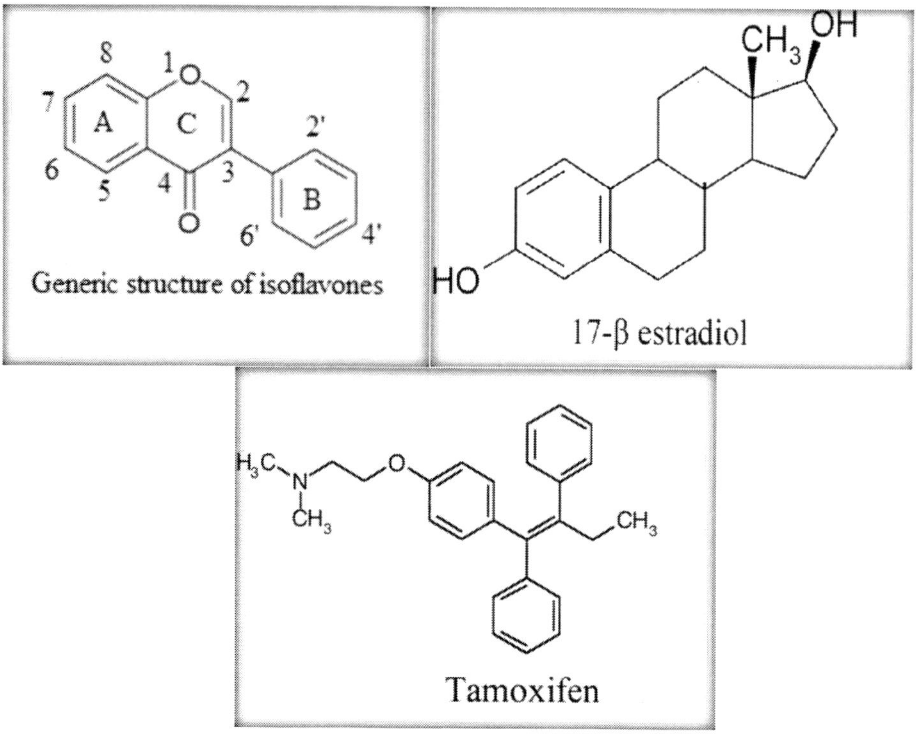

Generic structure of isoflavones

17-β estradiol

Tamoxifen

Figure 4. Chemical structure of isoflavones, 17β-estradiol and tamoxifen.

The estrogenic properties of the phytoestrogens were observed in the 1940s, when Australian sheep farmers noticed that their sheep were infertile. The ewes grazed on subterranean or red clover and their infertility was attributed to the presence of isoflavones like genistein and daidzein, along with their precursors, biochanin A and formononetin. Several estrogenic effects were found on different tissues including altered cell differentiation in the cervix leading to abnormal glands that produced a more hostile cervical mucus; weight gain of the uterus; ovulation abnormalities due to inhibition of follicular development or follicular degeneration; altered central nervous system/pituitary function, and altered sexual behavior [57]. Isoflavone consumption has also been associated with infertility in cattle [58], California quail [59], and captive cheetahs fed commercial diets that contained high levels of genistein and daidzein, suggesting that the presence of dietary isoflavones were responsible for their infertility [60,61].

Isoflavones

Isoflavones, or isoflavonoids, are the most common type of phytoestrogens although with a restricted distribution in the plant kingdom. They are mostly limited to the subfamily Papilionoideae of the Leguminosae [62] and are found in highest amounts in soybeans and soyfoods (like miso and tofu), being also present albeit at lower concentrations in other beans, fruits, vegetables and legumes [63,64]. Soy foods generally contain 1.2 – 3.3 mg isoflavones/g dry weight, with the precise amount depending on a number of factors that include the type of soy food as well as soybean variety, harvest year, geographical location, environmental conditions, and extent of industrial processing of the soybeans [65,66].

The major isoflavones present in soy foods are genistein, daidzein and, to a lesser extent, glycitein. These compounds are, in fact, hydrolysis products of glycosides in plants, termed genistin, daidzin and glycitin, which, after oral ingestion, are extensively biotransformed in the intestine by the action of bacterial enzymes [5]. The extent of intestinal bacterial metabolism will therefore have an effect on the bioavailability of dietary estrogens and, consequently, be expected to influence the potential for physiologic effects [5]. Another source of isoflavones is clover. It contains high concentrations of formononetin, which is a precursor of daidzein, and biochanin A, which is a precursor of genistein [63].

The flavonoids are a large chemical class of substances, in which the isoflavones are included, that are formed through the prenylpropanoid-acetate biodemal pathway via chalcone synthase and condensation reactions with malonyl CoA. The first step in flavonoid biosynthesis is the condensation of *p*-coumaroy CoA with three molecules of malonyl CoA to give chalconaringenin (= naringenin chalcone), a reaction catalyzed by chalcone synthase [67] (Figure 5).

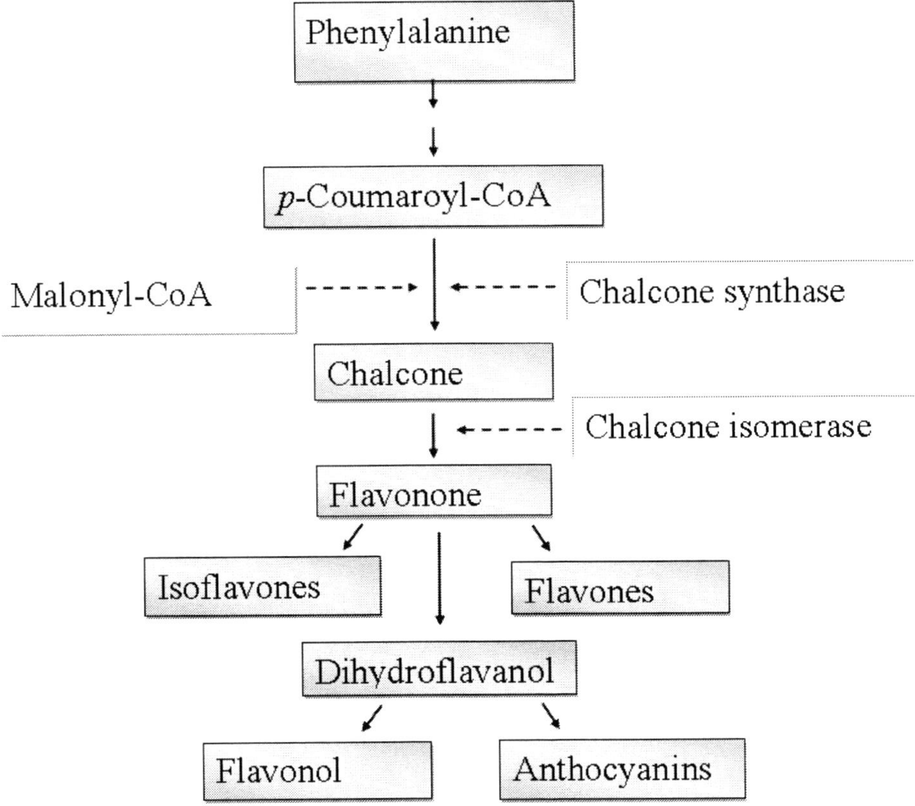

Figure 5. Scheme of general flavonoid pathway.

The isoflavones possess a 3-prenylchroman skeleton that is biogenetically derived from the 2-prenylchroman skeleton of the flavonoids [68]. The compounds of the isoflavones share a common structure, with genistein (4’, 5, 7 – trihydroxyisoflavone) having the important –

OH groups at positions 7 and 4', and biochantin A, a methoxy group at position 4' [9]. Small differences in the structure of the isoflavones may affect their activity. For example, daidzein differs from genistein only by the absence of a hydroxyl group on the A ring, but this difference greatly affects the functions of these two molecules [36].

The chemical composition of the dietary isoflavone may be a key determinant of its bioavailability and extent of biotransformation. The absence of isoflavones in plasma occurs 24 hours after oral intake [5]. The peak concentration range in women was 300 – 3200 nmol/L (80 – 800 ng/mL), but the time to reach this concentration was between 6 and 8 h after ingestion, and the half-life of plasma disappearance was at 7.9 h. Similar pharmacokinetics were obtained for adult men. These are quite high concentrations of isoflavones when compared with the endogenous plasma concentrations of estradiol throughout life, and despite the relatively weaker estrogenic activity, it is probable that the concentrations are sufficient to account for the many *in vivo* and *in vitro* biological effects that have been attributed to isoflavones [5].

The large number of phytoestrogens is introduced into the diet as inactive compounds. After their ingestion, a complex enzymatic conversion occurs in the gastrointestinal tract resulting in the formation of compounds with a steroidal structure similar to estrogens [18]. Considering the isoflavone, they can be found as glucosides or as aglycones. The glucosides are readily hydrolyzed in the gut to their aglycones, which are then easily transported across intestinal epithelial cells [69]. In the intestine, bacteria take an important part in the absorption and metabolism of isoflavones by producing the enzymes, termed glucosidases, which are responsible for metabolizing the glycosidic isoflavones (genistin, daidzin and glycitin) to their corresponding aglycones (genistein, daidzein and glycitein) [5,17] (Figure 6). However, before absorption occurs, intestinal bacteria may further metabolize the isoflavone aglycones to isoflavone metabolites. This metabolic process particularly involves genistein conversion to *p*-ethyl phenol in the colon (Figure 7), and daidzein to dihydrodaidzein, equol and/or O-desmethylangolensin (O-DMA) (Figure 8), all of which may also be absorbed [5,17]. The physiological effects of these metabolites may differ from those of the primary isoflavones, like equol and O-DMA, which are more powerful antioxidants than genistein and daidzein [70]. Daidzein, genistein, equol and O-DMA are the major isoflavones detected in blood and urine of animals and humans. In addition, dihydrodaidzein, *p*-ethyl phenol and glycitein (Figure 9) have been detected in human plasma and urine [71,72].

After absorption, isoflavones go through hepatic conjugation to glucuronic acid or sulphate to produce forms that are measured in biological fluids [5,17,73]. Similar to endogenous steroids, they undergo enterohepatic circulation following deconjugation in the intestine and re-absorption or excretion in the feces [5,17]; or after enterohepatic circulation, they may be excreted in bile or deconjugated, reabsorbed, reconjugated, and excreted in the urine. The efficiency of conjugation of isoflavones is high and consequently the proportion of circulating free isoflavones is small [5]. Studies in humans have shown that serum and urinary concentrations of isoflavones increase in accordance with the amount consumed, indicating that absorption occurs in a dose-dependent manner [74]. Moreover, studies using a controlled quantity of soy have shown a variable individual metabolic response with up to a 1000-fold variation in isoflavone excretion [16]. It is supposed that the composition of the intestinal flora, intestinal transit time, and variation in redox potential of the colon might influence the variability observed in humans [75]. This metabolic variability of the phytoestrogens is clinically important as it may influence biological responses. In healthy humans, plasma concentrations of isoflavones are usually less than 40 nmol/L [76] when consuming soy-free diets. On the other hand, plasma concentrations of isoflavones increase markedly in the micromolar range after single ingestion of soybean milk [77], a soy meal [78] or baked soybean powder [79]

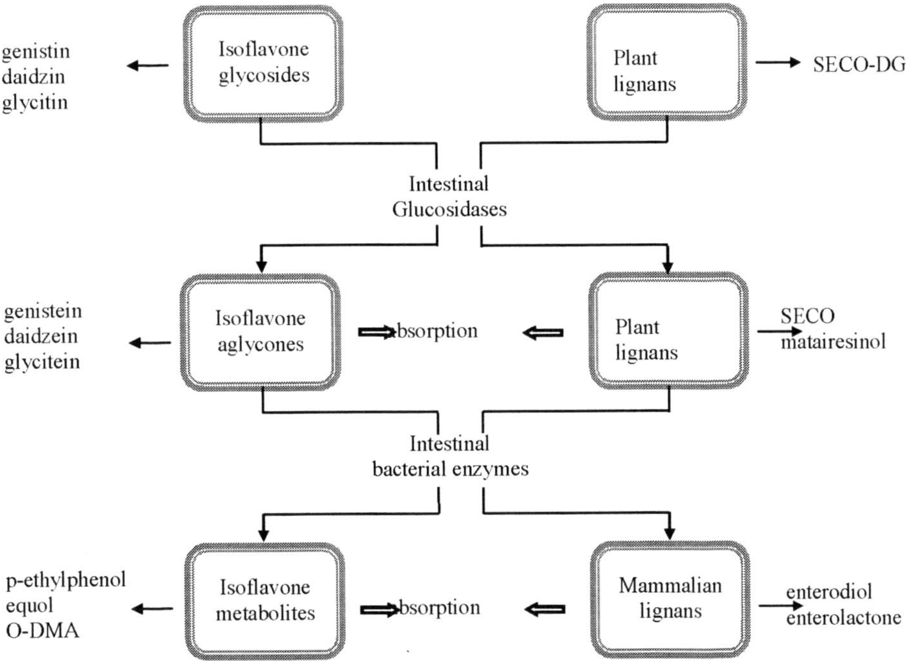

Figure 6. Schematic presentation of the absorption and metabolism of isoflavones and lignans. O-DMA = O-desmethylangolensin; SECO-DG = secoisalariciresinol-diglucoside; SECO = secoisalariciresinol. (Based on reference 110).

Figure 7. The metabolic pathway of genistin to its aglycone and metabolic product.

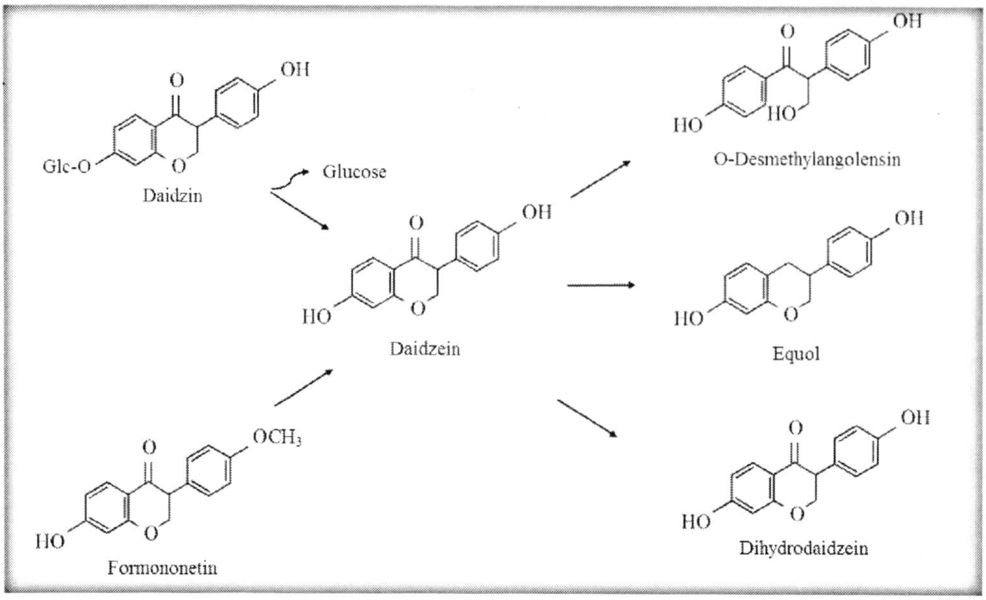

Figure 8. The metabolic pathway of daidzin to its aglycone and metabolic products.

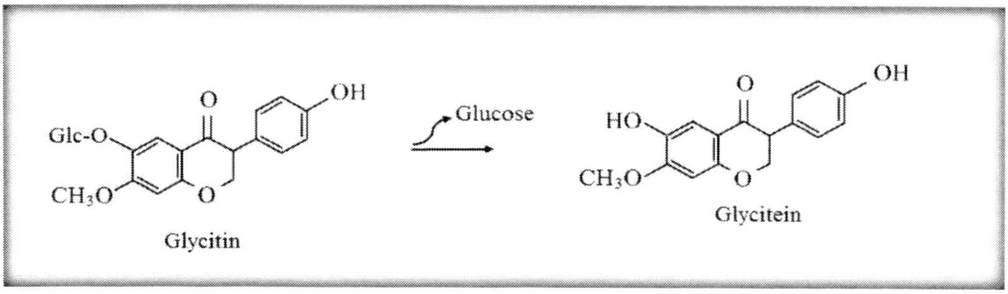

Figure 9. The metabolic pathway of glycitin to its aglycone.

In non-fermented soy foods, the isoflavones are mostly present as betaglucosides, some of which are esterified with malonic [80,81] or acetic acid [81,82]. In fermented soy foods, such as miso, soybean paste or tempeh, the isoflavones are largely unconjugated [83]. These chemical differences have been shown to alter the isoflavones` pharmacokinetics in the rat [84] and therefore probably in humans. The unconjugated isoflavones, like many common drugs, are absorbed from the upper small intestine in the rat [85], whereas the isoflavone glycoside conjugates are absorbed more slowly, indicating that their hydrolysis occurs at more distal sites in the intestine in relation to the unconjugated isoflavones [78,84].

Dietary phytoestrogens are weakly estrogenic (10^{-2} to 10^{-3}-fold, depending on the system examined) when compared with estradiol (E_2) or estrone, the main circulating estrogens of most mammals [86,87,88]. The preferential binding of nonsteroidal estrogen to the ERβ receptor suggests that they may exert their actions through distinct pathways from those of classical steroidal estrogens [5]. In addition, the lower binding rate of several phytoestrogens to serum proteins would indicate an increase in the numbers of molecules available for receptor occupancy [89]. Of the isoflavones, genistein and daidzein are thought to exert the most potent estrogenic effect. Numerous studies have shown that the isoflavones, as well as the lignans and the fungal estrogens, compete effectively with E_2 for binding ER-rich cells [12]. The phenolic B-ring of the isoflavones confers binding to estrogen receptors (ERβ > ERα, although at a lower affinity relative to estradiol) and acts like SERMS [25,53,90]. Genistein has one-third the potency of E_2 when it binds to ERβ, and one thousandth of the potency of E_2 when it binds to ERα as determined by expression of luciferase reporter gene construct in kidney cells cotransfected with ERα and ERβ [58].

Genistein can induce similar responses in prostate, endometrial, ovarian, breast, bone tissues and cell lines as estradiol [91,92,93,94], but it can also act as an estrogen antagonist in some tissues. When neonatal prepubertal Sprague-Dawley rats were treated with estrogens, an increase in the number of terminal end buds and cell proliferation in mammary tissue was observed [72]; however, after treatment with genistein, a reduction in the number of terminal end buds and cell proliferation was observed [95]. Other studies showed that genistein inhibited the development and growth of chemically induced tumors in the breast [96] and prostate [97] of mice and rats. It is believed that genistein and other isoflavones exert this effect through induction of a signal transduction pathway leading to apoptosis [98]. In the rat uterus, genistein down-regulated ERα and up-regulated progesterone receptor, suggesting signaling via ER [99], while in castrated mice, it up-regulated the estrogen-responsive gene (c-fos), in the prostate, suggesting an estrogenic mechanism [100,101].

LIGNANS

Lignans are more widely distributed in plants, being highly concentrated in flaxseed, but also found in oilseeds, seaweed, legumes, seeds, vegetables (especially carrots, spinach, broccoli and cauliflower), whole grains and tea. Fruits have low levels of lignans with the exception of strawberries and cranberries [102]. Flaxseed generally contains 0.96 – 3.15 μmol lignans/g, with the precise amount depending on factors such as variety, growing location and harvest year [103]. The term lignan is used for a diverse class of phenylpropanoid dimers and oligomers. The two main lignan dimers, that are not estrogenic by themselves, are termed secoisolariciresinol-diglucoside (SECO-DG) and matairesinol (MAT), which serve as precursors to the mammalian lignans, enterodiol (2,3-bis(3-hydroxybenzyl)-butane-1,4-diol) and enterolactone (trans-3,4-bis(3-hydroxybenzyl)-dihydro-2-furanone) respectively [5,103] (Figure 10). Lignans` precursors occur in grains close to the outer fiber-containing layer, a part that modern Western milling techniques usually eliminate. As a result, phytoestrogens are seldom present in cereal diets in Western countries [18]. Other recently discovered lignans are pinoresinol and lariciresinol [104].

Lignans present different biological activities in animal and *in vitro* systems, including anticarcinogenic, antioxidant, weakly estrogenic and antiestrogenic properties, and inhibition of enzymes involved in sex hormones metabolism [56,105,106]. For instance, enterodiol inhibits 5α-reductase and 17β-hydroxysteroid dehydrogenase in human tissues. Therefore it is suspected that lignans may reduce the plasma levels of free estradiol and testosterone and may affect the development of hormone dependent diseases. The two major mammalian lignans (enterodiol and enterolactone) have been found in human plasma, saliva, feces, semen and prostatic fluid [56].

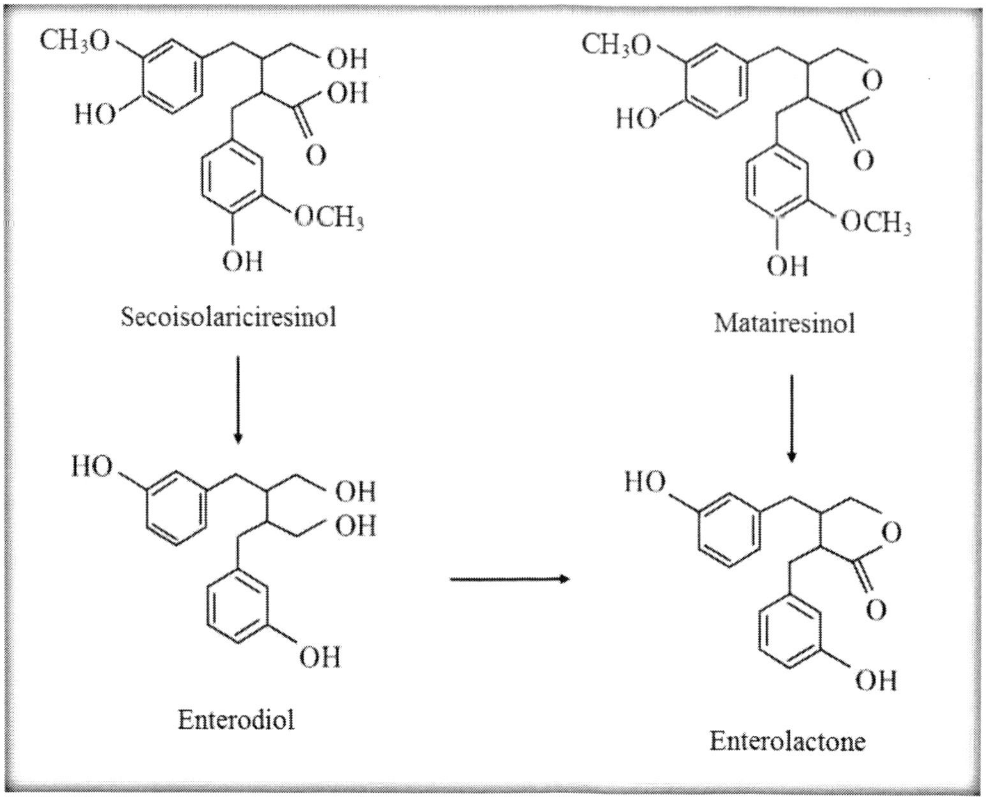

Figure 10. Chemical structure of lignans.

The absorption and metabolism of lignans are similar to that of isoflavones. The conversion occurs by gut microflora and the mammalian lignans are readily absorbed. After oral ingestion, the lignan SECO-DG is metabolized to SECO through hydrolysis of the sugar moieties by intestinal bacteria, followed by dehydroxylation and demethylation to enterodiol, reactions which are also catalyzed by bacterial enzymes. Enterodiol may be absorbed or further oxidized irreversibly to enterolactone, which is then absorbed. Intestinal bacterial enzymes also catalyze dehydroxylation and demethylation reactions to convert MAT to enterolactone, which can then be absorbed. Once mammalian lignans are absorbed, they undergo conjugation to glucuronic acid or sulphate and enter the circulation. After this, they may be excreted in the urine or undergo enterohepatic circulation [107]. Studies have shown that urinary and plasma concentrations of the mammalian lignans increase significantly in a dose-dependent manner following consumption of lignan-rich flaxseed [108,109]. A cross-sectional study reported a positive association between plasma enterolactone and consumption of vegetables and fibers, however, positive associations were also observed with alcohol and caffeine intake, raising concern regarding the specificity of plasma enterolactone as a biomarker for exposure to lignan-containing foods [110,111].

COUMESTANS

Coumestans are found in large quantities in plants, but only a small number have shown estrogenic activity, namely coumestrol (Figure 11) and 4' methoxy coumestrol [35]. Although they are much less commonly consumed, coumestrol is mainly found in legumes and vegetables, particularly soy sprouts and alfalfa sprouts [112]. Clover and soybean sprouts are reported to have the highest concentration, 28 and 7 mg/ 100 g dry weight, respectively, whereas mature soybeans only have 0.12 mg/100 g dry weight [113,114]. The estimated daily intake of coumestrol by US women is 0.6 μg/day, a much lower value than that of isoflavone intake (154 μg/day) [115]. Coumestrol binds twice as well to ERβ than to ERα and binds to ERβ with twice the affinity of estradiol [56,116].

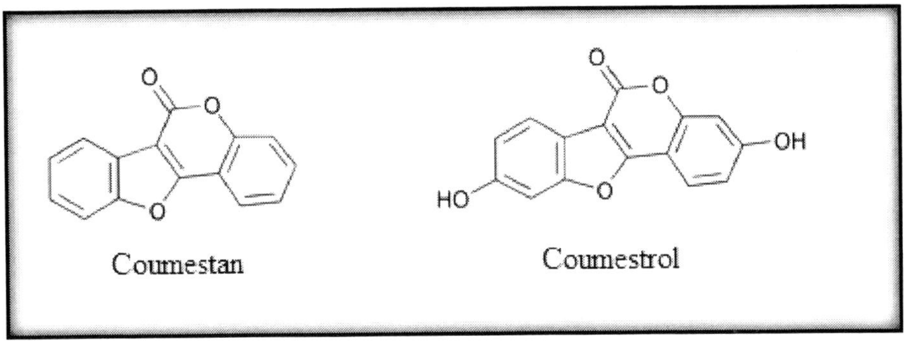

Coumestan Coumestrol

Figure 11. Chemical structure of coumestan and coumestrol.

PRENYLFLAVONOIDS

The prenyflavonoids are primarily represented by 8-prenylnaringenin, which is known to be the most potent phytoestrogen (Figure 12). Beer is the most important dietary source of 8-prenylnaringenin, and its presence in this beverage is due to isomerization of desmethylxanthohumol in the brew kettle. Other prenylated flavonoids are xanthohumol and isoxanthohumol, also found in beer [67].

Although it is a much weaker estrogen than 17β estradiol (< 1%), the presence of 8-prenylnaringenin in beer has resulted in concerns as to the estrogenic effects of beer consumption. Most hopped beers contain less than 100 μg of 8-prenylnaringenin/L, and the highest amount concentration measured in microbrew beer was 240 μg/L [67]. In ovariectomized mice treated with 8-prenylnaringenin, the minimal concentration to cause estrogenic effect, i.e. to produce a significant increase in vaginal mitosis was 100 μg/mL drinking water. No effect on uterine growth was observed at this concentration [117], which is about 400 – 1000-fold higher than the 8-prenylnaringenin concentration in beer. Based on this information, it seems that it is safe to assume that human exposure to phytoestrogens through beer consumption is not a health threat.

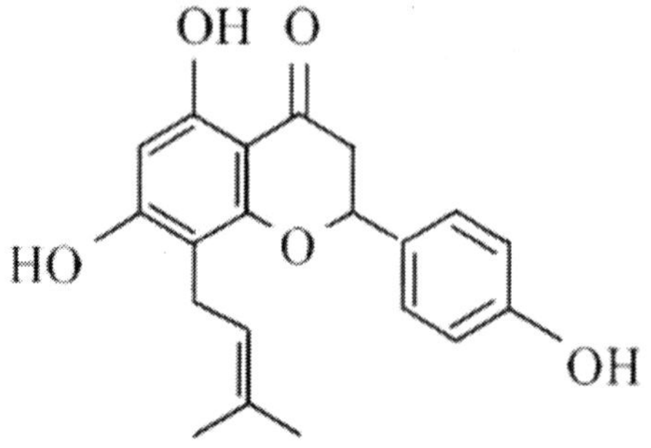

Figure 12. Chemical structure of 8-prenylnaringenin.

The daily intake of prenylflavonoids, approximately 0.14 mg, is relatively small compared to total polyphenols from beer (42 mg as catechin equivalents per day) [118], indicating that the prenyflavonoids play a minor role to the antioxidant properties of beer. Prenylflavonoids differ from other polyphenols, such as flavonol glycosides and phenolic acids, in that they are more lipophilic and, consequently, may be more effective antioxidants at lipophilic surfaces such as membranes and low-density lipoprotein [119].

A competition assay with recombinant human ERα and ERβ from cytosolic SF-9 cell extracts has shown that 8-prenylnaringenin has a high affinity and strong selectivity for the ERα. In comparison with genistein, 8-prenylnaringenin was 100 times more potent ERα agonist but a much weaker agonist of ERβ in the estradiol-competition assay for receptor binding. It is believed that, to date, 8-prenylnaringenin is the strongest plant-derived ERα receptor agonist identified. An *in vivo* study has shown that the estrogenic potency of 8-prenylnaringenin was about 20.000-fold lower when compared to 17β estradiol using uterine and vagina growth assays [120].

STILLBENES

Stillbenes, like the flavonoids, are produced through the prenypropanoid-acetate pathway. The main dietary source of stillbenes is resveratrol (Figure 13) found in red wine and peanuts although it is also synthesized by a variety of plant species in response to injury, ultra-violet irradiation and fungal attack [121]. There are two isomers of resveratrol, *cis* and *trans*, but only the *trans* form is estrogenic [122]. Resveratrol is only found in the skin of the grape, resulting in low levels of *trans*-resveratrol in white wine. Its content in wine depends on cultivar, geographic location, season, oenological practices and presence of Botrital fungus [123,124,125]. The longer the fermentation time the more *trans*-resveratrol will be in the wine. The level of *trans*-resveratrol in red wines, fermented with skins, can be as high as 14.5 mg/ L [126], but the type of post harvest processing has a large effect on the resveratrol content. For instance, the hot extraction method used for obtaining purple grape juice, as opposed to the cold pressing of grapes for obtaining the white grape juice, provide a higher concentration of resveratrol [124]. This is also true for peanuts, with boiled peanuts containing more resveratrol than peanut butter and roasted peanuts [127]. In peanuts, resveratrol is found throughout the nut; however, on a weight basis the seed-coat has the highest levels [128]. The smaller peanuts show higher resveratrol content, but as the peanuts mature the resveratrol content in the nut declines [127].

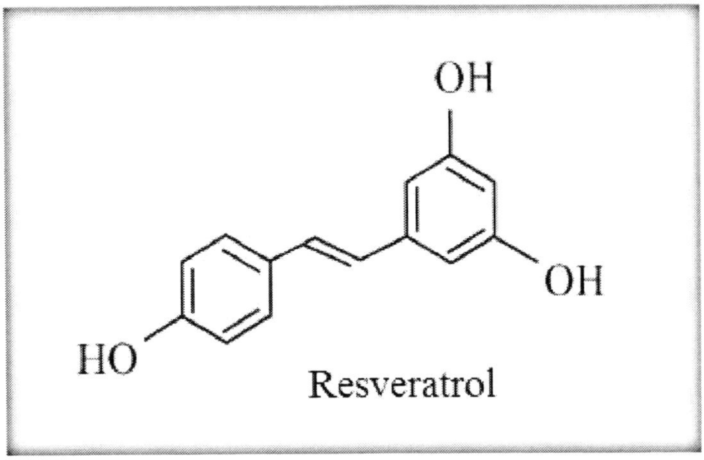

OH
OH
HO
Resveratrol

Figure 13. Chemical structure of stilbene resveratrol.

Other sources of stillbene have been reported in the literature. They include the roots of *Polygonum cuspidatum* (Japonese knotweed or Mexican bamboo) used in China for its therapeutic properties, with resveratrol levels in the dried roots being as high as 377 mg/100 g dry weight [129]. Resveratrol has also been isolated from several grass species [130], pine bark [131], ivy and lilies [124].

Resveratrol physiological levels can be obtained through drinking red wine. It shows a greater capacity to activate the ERβ than ERα [132]. Its agonistic and antagonistic activity has been shown in MCF-7 cells and the hamster ovarian cell line (CHO-K1) transfected with human ERα and ERβ [133,134]. Resveratrol is considered to be an effective radical-scavenger, suggesting that it acts as a natural antioxidant against oxidative DNA damage [135,136]. It also exhibits anti-inflammatory effects and antimutagenic and anticarcinogenic activity [121]

Effects on Human Health

Owing to the ability of the phytoestrogens to exert estrogenic activity, there has been considerable scientific interest on their potential beneficial health effects, particularly as anticarcinogens, cardioprotectants, in osteoporosis, and as alternatives to hormone replacement therapy in menopause. The incidence and mortality of hormone-related cancers such as breast, endometrial and prostate cancer vary around the world, being higher in the

Western countries than in Asian countries. These differences are being related to variations in the diet, and one such example is the intake of phytoestrogens.

Cancer

Numerous epidemiological studies have shown that people who consume high amounts of isoflavones in their diet have lower rates of several cancers, including breast, prostate and colon cancer. The idea that soy may have a preventive role in cancer came from a broader research of the reasons for the large worldwide variation in cancer incidence and death. Americans and western Europeans are at much higher risk of breast cancer and prostate cancer than are peoples from the Asian countries (137). In Asia, where the consumption of phytoestrogen is much higher than that of the Western countries, the rates of these cancers are low (Figure 14). It is estimated that the average American diet contains less than 1 - 3g per day of isoflavone, whereas the average Japanese diet contains 30 - 50 g per day [138]. Immigrants from Asia living in the US have an increased risk of cancer compared with the length of stay in the US and the exposure to the North American diet [139]. For women, this change occurs mostly in the next generation, whereas for men, the increase in prostate cancer risk occurs in the same generation [140].

The antiproliferative properties of isoflavones suggest that these compounds may inhibit the cell cycle or induce apoptosis. It has been reported that concentrations of 5 – 20 μg/mL genistein caused cell cycle arrest at the G1/S and G2/M phases in the human myelogenous leukemia HL-60 and the lymphocytic leukemia MOLT-4 cell line [141], and concentrations of up to 60μM genistein arrested human gastric cancer cells at G2/M [142]. In addition to genistein, genistin, daidzein and biochanin A at 0 - 50 μM inhibited murine and human cancer cells growth by inducing cell cycle arrest and apoptosis [143].

Apoptosis is an effective beneficial antitumor process since it removes damaged or abnormally functioning cells from tissues. At high concentrations, genistein is reported to induce apoptosis through mitochondrial dependent pathways. For example, 50 μM (24 h) genistein induced cytochrome c release, caspase- 3 activation, nuclear condensation and DNA fragmentation by reducing the mitochondrial membrane potential [144]. Lower concentrations of genistein (15 - 60 μM, 24 h) induced apoptosis in murine T-cell lymphoma cell lines via mitochondrial depolarization and activation of caspases 3 and 9 [145]. Genistein has also been suggested to induce apoptosis by inhibiting DNA topoisomerase II, thereby preventing ligation of DNA double strand breaks [146].

The isoflavones might protect against cancer and/or heart disease through inhibition of oxidative damage [147], as DNA oxidation is probably an important cause of mutations that potentially can be reduced by dietary antioxidants. The isoflavones serve as derivatives of conjugated ring structures and hydroxyl groups that may function as antioxidants in *in vitro* cell culture or cell free systems by scavenging superoxide anion [148], lipid peroxy-radicals [149], singlet oxygen [150], and/or stabilizing free radicals involved in oxidative processes through hydrogenation or complexing with oxidizing species [151].

	Genistein	Daidzein	Biochanin A	Formononetine	Coumestrol	SECO	Metairesinol
Whole grain	---	---	---	---	---	32.9	2.6
Barley	7.7	14.0	---	---	---	58.0	0
Wheat bran	6.9	3.5	---	---	---	110	---
Soy flower	93 900	67 400	70	30	---	130	---
Soybean	26 800 – 102 500	10 500 – 85 000	---	18.0 – 121.0	---	13.0 – 273.0	tr
Tofu	21 300	7 600	---	---	---	---	---
Miso	14 500	7 300	---	---	---	---	---
Soy drink	2 100	700	---	---	---	---	---
Soy Milk	310	30	---	---	---	---	---
Bean	365	9.7	14	7.5	1.8	172	0.25
Chick peas	69.0 - 214.0	11.0 – 192.0	838	215	5	7.0 – 8.0	0
Flaxseed	---	---	---	---	---	369 900	1087
Sunflower seed	13.9	8.0	---	---	---	610	0
Clover seed	323	178	381	1270	5.3	13	3.8
Alfalfa sprouts	5.0	62.0	124.0	4090	45.0	33.0	0
Peanut	64.0	58.0	31	---	---	298.0	tr
Strawberry	tr	tr	---	---	---	3718	22.5
Apple	tr	12.4	---	---	---	5.0	0
Banana	0	0	---	---	---	10.0	0
Orange	0	0	---	---	---	76.8	0
Avocado	0	0	---	---	---	76.7	16.0
Guava	0	0	---	---	---	699.7	tr
Broccoli	8.0	6.0	---	---	---	414.0	23.0
Cauliflower	9.0	5.0	---	---	---	97.0	tr
Garlic	tr	tr	---	---	---	379.0	3.6
Carrot	0	0	---	---	---	192.0	3.0
China Green tea*	tr	tr	---	---	---	2890	195.0
China Black tea*	0	0	---	---	---	1050	90.0

Figure 14. Phytoestrogen content inselected food items.

BREAST CANCER

Cancer chemoprevention is different from cancer treatment because it is aimed at modulation of pathways that are relevant to carcinogenesis with the purpose of inhibiting, reversing, or retarding cellular hyperproliferation. It is targeted at the initiation, promotion and progression stages of carcinogenesis and requires long-term exposure to non-toxic nutrients, food supplements or pharmacological agents in order to prevent the development of malignancies [67,68].

There are two basic ways of cancer prevention -- primary and secondary. In primary prevention, tumors do not appear clinically as a result of preventive measures that reduce the risk of genetic damage, the precursor to cancer at the initiation stage later in life, or stop the mutated cancer cell from growing abnormally and becoming an observable tumor. The secondary prevention aims at the prevention of another cancer from occurring, as after the removal of a breast tumor, a patient may develop a new primary tumor at a site different from that of the first one, usually the other breast. Since most breast tumors express functional estrogen receptors, a strategy for preventing the growth of tumor cells remaining after surgical removal of the mammary tumor is the anti-estrogen tamoxifen. As tamoxifen is tumouristatic rather than tumouricidal, it was later realized that it had a potential role as a prophylactic, primary and/or secondary prevention agent for breast cancer [152]. In relation to the natural occurring substances, one hypothesis regarding the breast cancer protective effect of phytoestrogens was associated with the prolonged follicular phase. In breast epithelial cells, proliferation is low during the follicular phase and increased two fold in the luteal phase. Women with longer cycles secondary to increased follicular phase therefore have decreased cell division. Less breast cell proliferation reduces the risk of dysplasia resulting in breast cancer [57,153].

The protective effect of phytoestrogens on cancer may also be due to their role in lowering circulating levels of unconjugated sex hormones. As mentioned earlier, estrogens primarily circulate as inactive conjugates of sex hormone binding globulin (SHBG) or albumin [35]. In postmenopausal women, dietary supplementation with soy isoflavones or lignans increased the levels of SHBG [154,155,156,157], and thereby lowered the serum levels of free estradiol [157,158]. In menstruating women, however, the effect of phytoestrogens on estrogen levels is inconsistent because most supplementation studies failed to report any difference in menstrual cycle length or hormone levels with soy supplementation [159,160,161,162].

The consumption of isoflavone and flaxseed significantly changed the urinary excretion of genotoxic estrogen metabolites related to decreased breast cancer risk. Three major pathways, leading to different metabolites, can metabolize estradiol and estrone. Two of these pathways produce 16α-hydroxyestrogens and 4-hydroxyestrogens, which are known to be genotoxic [163], whereas the 2-hydroxyestrogen metabolites are believed to play a protective role on breast cancer. Following high ingestion of soy and flax, it was observed that the ratio of 2-hydroxyestrogens to 16α- hydroxyestrogens increased and this is considered to be an important breast cancer biomarker [164]. In addition, the level of serum SHBG has been negatively correlated with the urinary excretion of 16α- hydroxyestrogens [154]. Other potential mechanisms by which phytoestrogens may protect against breast cancer risk are independent of the estrogen receptor and include inhibition of enzymes involved in the synthesis of steroid hormones, including aromatase and 17β-hydroxysteroid dehydrogenase [165,166].

Urinary excretion of phytoestrogen metabolites can be used as a biomarker for the intake and bioavailability of the phytoestrogens. The data available indicates a lower risk of breast cancer in women with increased excretion of phytoestrogens, and, in case control study, breast cancer patients had significantly lower excretion of isoflavones and lignans than healthy subjects [167].

Neonatal exposure to genistein has been shown to prevent mammary cell tumor in rats, possibly through nonestrogenic mechanisms. The mechanisms proposed to explain the protective effect of genistein against cancer include induction of differentiation, inhibition of DNA topoisomerases, inhibition of specific cell cycle events, induction of apoptosis, inhibition of angiogenesis, antioxidant activity (due to the phenolic ring structures), inhibiting production of hydrogen peroxidase, and inhibition of protein tyrosine kinase (PTK), which is associated with cellular receptors for several growth factors such as epidermal growth factor (EGF), platelet-derived growth factor (PDGF), fibroblast growth factor (FGF) and mononuclear phagocyte growth factor [168,169].

Soybeans have been shown to contain protease inhibitors and there are at least two different inhibitors measurable in soybeans: Kunitz trypsin inhibitor and Bowman-Birk protease inhibitor (BBI). These may help to prevent conversion of normal cells to malignant ones, even very late in the process of carcinogenesis, but have no effect on cancer cells. BBI has been shown to prevent experimentally induced colon, oral, lung, liver, and esophageal cancer in animals. It is thought that this anticarcinogenic effect comes from its inhibition of chymotrypsin; however, there are conflicting data on the heat stability of these protease inhibitors and thus on their activity in foods [170]. Another mechanism of action of isoflavones is through modulation of gene transcription. For example, genistein modulates transcriptional regulation of transforming growth factor-β1 (TGF-β1), a multifunctional cytokine involved in cell proliferation and differentiation and extracellular matrix (ECM) synthesis [171]. Isoflavones may also inhibit PTK-dependent activation of transcription factors, such as nuclear factor-KB (NF-KB) and activator protein-l which are important signaling molecules involved in the response of cells to inflammation and oxidative stress [172,173,174,175]

The timing of exposure to the anticancer components of soy may be crucial. In an experimental model of induced breast cancer using neonatal and pre-pubertal rats, genistein not only suppressed by 50% the number of mammary tumors observed over a 6-month period [95,176], but it also delayed the appearance of the tumors. Genistein was initially administered in high doses by injection; however, similar degrees of protection were later obtained by placing genistein in the feed (0.25 g/kg) of the mother, whereby the offspring were exposed to genistein from the diet from conception through to day 21 postpartum [96].

In addition to the isoflavones, lignans and prenylflavonoids have also been shown to contribute towards reducing breast cancer risk. Vegetarians, who have a low risk of breast cancer relative to that of the omnivores, also have a significantly higher dietary intake and urinary excretion of lignans [177]. High plasma concentration of the mammalian lignan enterolactone is correlated with a reduced risk of breast cancer [178,179]. Furthermore, urinary enterodiol [180] and enterolactone [180,181] are significantly lower in post-menopausal breast cancer patients relative to controls. Consistent with these data are studies showing that consumption of flax or purified lignans reduces the risk of mammary cancer in rodents [182]. Experimental data have indicated that hop prenylflavonoids have potential as cancer chemopreventive agents by interfering with a variety of cellular mechanisms at low micromolar concentrations, such as inhibition of metabolic activation of procarcinogens, induction of carcinogen-detoxifying enzymes, and inhibition of tumor growth by inhibiting inflammatory signals and angiogenesis.

Despite the supportive data regarding cancer chemoprevention of the phytoestrogens, the epidemiological studies relating breast cancer risk to intake and/or urinary excretion of these compounds have not been consistent. For instance, there was no observed relationship between consumption of soy foods and breast cancer risk in a diet and cancer case–control study in Chinese and Japanese women [183,184]. Similarly, in a case–control study involving multiethnic American women, breast cancer risk was also not affected by either isoflavone or lignan intake [185], and no significant associations were observed between urinary genistein or enterolactone and breast cancer risk in a prospective study of post-menopausal Dutch women [186]. Finally, in Finland, a case control study found no correlation between serum enterolactone concentration and the risk of developing breast cancer in pre or post menopausal women [187].

PROSTATE CANCER

Prostate cancer is the most commonly diagnosed malignancy in man in the world, with incidence rates varying considerably between populations and countries. The lowest rates, however, are usually observed in Asia, and some migration studies have shown that Japanese migrants to the US present increased incidence of prostate cancer [140]. The high intake of isoflavones by Asian men is evident by the high concentrations of soy isoflavones found in their blood, urine and prostatic fluid, compared with their European and Western counterparts [110].

Prostate cancer is a major health concern and treatment is principally based on its hormone dependence, as this type of cancer is dependent on androgen for development, growth, and survival [188]. Therefore, substances that alter hormone action can have substantial biological effects on prostate cancer development and progression.

The biological activity of androgens is triggered by the activation of the androgen receptor (AR) when binding to its specific ligand [188,189]. Like the estrogen receptors, AR is a member of the nuclear receptor superfamily that functions as a ligand-dependent transcription factor, which, before ligand activation, is present in the cell in an inactive state through association with inhibitory heat shock proteins. Testosterone is the primary AR ligand in serum, but is converted to the more potent dihydrotestosterone (DHT) in prostatic epithelial or adenocarcinoma cells through the action of the enzyme 5α reductase, [190,191]. DHT is a high-affinity ligand for AR, and following DHT binding, a series of events occur, including displacement of heat shock proteins, receptor homodimerization, and rapid translocation to the nucleus. Upon activation, AR then binds to specific DNA sequences, termed androgen responsive elements (AREs), and initiate gene transcription [188]. This results in a myriad of biological response that may involve differentiation, secretion, and cellular proliferation. In prostate cancer, the AR function is required for the development and survival of the tumor [188,192], and the androgen depravation therapy is designed to block AR function [193] either through preventing androgen synthesis or through the use of direct AR antagonists. At the cellular level, the androgen depravation therapy triggers a cascade of events leading to either programmed cell death or cell cycle arrest [193,194].

A group of environmental and industrial substances are classified as endocrine disrupting compounds (EDCs) due to their ability to simulate, alter, or block endogenous hormone action. Some of these compounds may act through the activation of nuclear hormone receptors and may present estrogenic activity, being referred to as estrogenic endocrine disrupting compounds (EEDCs) [195]. Similar to endogenous estrogens, several EEDCs have been hypothesized to alter androgen action or prostate growth and development [196]. These agents can be classified into naturally occurring compounds such as phytoestrogens or synthetically generated agents.

A considerable amount of evidence has revealed the potential beneficial effects of phytoestrogens on prostate cancer development. However, these effects may be influenced by the discrete biological activities of metabolites and by variations in the metabolic profiles of different individuals. Several studies have pointed out that phytoestrogens do not act via a single mechanism of action, nor do different phytoestrogens produce the same profile of biological responses. The protective effect of phytoestrogens has been investigated in animal models and in humans. Studies in humans have described an association between diets rich in soy, lignan, or rye and decreased prostate cancer risk [197]. In rats or mice, a diet rich in phytoestrogens resulted in reduced incidence of prostate tumors [198,199], and the inhibition of prostate tumor development was observed in rats fed an isoflavone-rich diet [200]. In a transgenic mouse model of prostate cancer, growth and metastatic potential of tumors was reduced as a result of genistein exposure [201]. The developing rat dorsolateral prostate showed reduced expression of AR, ERα and ERβ after oral exposure to genistein [202], revealing one possible mechanism behind the lower incidence of prostate cancer in populations consuming high levels of phytoestrogens. Resveratrol has also been examined in animal models of prostate cancer and has showed a chemopreventive activity [203].

Most prostate cancer studies have been conducted in animals fed with soy, pure isoflavones, or rye-bran. In particular, genistein has been shown to inhibit the growth of cancer cells through modulation of genes involved in the homeostatic control of the cell cycle and apoptosis. In this case, genistein inhibited activation of nuclear transcription factor, NF-kappaB and the Akt signaling pathway, both of which are known to maintain balance between cell survival and programmed cell death [204,205,206,207,208].

The consumption of phytoestrogen has also been implicated in the improvement of patients being treated for prostate cancer. Intake of an isoflavone-supplemented diet for a minimum of 3 months was shown to decrease the rise of PSA levels in both ADT-sensitive and ADT-refractory prostate cancers [209]. Phytoestrogens have also been shown to influence the steroid hormone axis, including the production, metabolism and biological activity of sex hormones, and the regulation of numerous intracellular steroid-metabolizing enzymes, including aromatase and 17β-hydroxysteroid dehydrogenase [6,8,165,166,210]. They may prevent prostate cancer via a dose dependent reduction of serum 17β estradiol and testosterone levels, as reported in Japanese men [211]. The presence of estrogen receptor β in the prostate [212], which binds genistein with an affinity similar to that of 17β estradiol, suggests that the effect of phytoestrogens may be mediated by this target [213]. In a pilot study, prostate cancer patients in a low-fat, phytoestrogen supplemented diet, presented a reduced testosterone and androgen levels [214]. In rats, soy-based diets significantly decreased circulating testosterone levels and prostate weights [215]. *In vitro* studies have also revealed that high concentrations of isoflavones and lignans inhibited the growth of prostate cancer cells in a dose dependant manner regardless of AR expression or hormone dependency [216,217,218].

Although a large number of evidence has suggested the protective action of phytoestrogens on prostate cancer, some studies have not been entirely supportive and information is still inconsistent. For instance, two case–control [219,220] and one prospective study [221] conducted in Asian countries were unable to detect significant associations between soy consumption and prostate cancer risk.

MENOPAUSE

Some women experience a decrease in the quality of life during menopause especially because of sleep deprivation, hot flushes, mood swings, forgetfulness and difficulty concentrating. In postmenopausal women, studies have tried to analyze whether dietary supplementation can alleviate menopausal symptoms, but the results have been inconclusive.

Several clinical studies have indicated that a dietary increase in phytoestrogens in addition to a regular diet [222] and phytoestrogen supplementation are associated with an increase in quality of life as well as a significant amelioration of the signs and symptoms of the climacteric syndrome [18,223]. However, several other studies could not demonstrate any clinical effects, especially with regard to hot flushes [224]. Many reasons may account for these discrepancies found in the various studies, which include differences in patient selection, phytoestrogen content, and symptom assessment, and variations in endogenous estrogen production, intestinal resorption, and activation of orally ingested isoflavones [225].

In regard to the climacteric syndrome, a large number of clinical and experimental data on the use of phytoestrogens for the alleviation of signs and symptoms of this condition have indicated that the efficacy of isoflavones is reduced in women with idiopathic premature ovarian failure, premature ovarian failure after cancer treatment, after long-term hormone therapy, and when isoflavone treatment is initiated more than 3 years after menopause [226]. On the other hand, phytoestrogens were found to be most effectively in early postmenopausal women with a natural menopause and mild to moderate climacteric syndrome. The literature indicates that phytoestrogen compounds used to treat women with climacteric syndrome should contain 40 to 100 mg of isoflavones consisting of variable combinations of different aglycans (i.e., genistein, daidzein, glycitein, formononetin, and biochanin A). A daily dose of 40 – 100 mg of isoflavones is toxicologically safe and is consistent with the dietary content of isoflavones in Asian countries [227].

Another factor that may partially interfere with the effect of a phytoestrogen diet is the individual age of the woman when starting the diet. Asian women usually begin consumption of phytoestrogens early in childhood, whereas the Western women tend to adopt a phytoestrogen-rich diet or phytoestrogen supplementation during perimenopause or postmenopause. This late consumption may then be too late to exert a protective effect on the breast. Phytoestrogens may induce differentiation of breast epithelium during early childhood and puberty, thus making the breast epithelium less sensitive to noxious agents such as chemical carcinogens [228].

Overall, it appears that there is some evidence to support a role for soy isoflavones in the management of menopausal hot flushes, but the effects specifically due to soy isoflavones are still considered modest. The use 60 g of isolated soy protein powder containing 40 g of proteins and 76 mg phytoestrogens, in the aglycone-active form, has been shown to halve the number of hot flushes in postmenopausal women in a double-blind, placebo-controlled trial [229]. Other studies in peri and postmenopausal women have reported decreased hot flushes following consumption of soy; however, in many studies, the results were not significantly greater than the decrease observed for the control group [230,231]. Despite this, a few studies have found significant decreases in number [229] or severity [232] of hot flushes within the soy treatment group beyond placebo.

The effect of lignans on hot flushes has not been widely investigated and the median daily intake of lignans has been estimated at 578μg in American postmenopausal women [115]. A study using flaxseed as a treatment group found that it reduced the frequency of hot flushes; however the results were not significantly different from the soy or wheat groups [230].

A few studies have investigated the effect of phytoestrogen supplementation on the endometrium of postmenopausal women. No effects were observed in the endometrial thickness as measured with transvaginal ultrasound [233,234], or in uterine artery pulsatility index [234]

OSTEOPOROSIS

Bone health and osteoporosis are two major concerns to postmenopausal women, considering that the decline in estrogen production that occurs during menopause is an important contributing factor. Estrogen is important for maintaining bone density because it regulates the formation and resorption of bone [46].

During menopause, the circulating levels of estradiol is lower, thereby more calcium is lost from the bone into blood plasma, leading to osteoporosis [235]. The hormone replacement therapy (HRT) was designed to prevent or lower the incidence of osteoporosis in postmenopausal women, however, due to the relatively few benefits of the hormone replacement therapy for quality of life reported by the Women`s Health Institute [236], other alternatives, such as a dietary alternative, are being searched. It is in this regard that the phytoestrogens are being investigated in relation to bone health. Observations of significantly lower numbers of hip fractures in Asian women, when compared to Western women, generated this interest in the potential relation between phytoestrogens and osteoporosis risk [237].

Several *in vitro*, animal and clinical studies have suggested that phytoestrogens, particularly the isoflavones, are somewhat effective in maintaining bone mineral density (BMD) in postmenopausal women [230,238,239,240]. Results from ovariectomized rodent models have demonstrated a beneficial effect of phytoestrogens on the retention of bone mass. Furthermore, they have indicated that a threshold dose of isoflavones needs to be consumed for a sufficiently long period of time, i.e. approximately 1 month to years depending on the species, before any measurable effects on bone mass and density can be observed. When animals consumed higher doses of isoflavones, the effect was no greater than that of the minimal threshold dose, and the higher doses may be less effective.

Isoflavones have been demonstrated to increase the synthesis of vitamin D in a number of nonrenal cell types, and, due to this mechanism, they are thought to contribute to bone mineral density [228,241]. In a double blind placebo controlled study of postmenopausal women, the daily administration of 54 mg genistein for a period of 12 months showed a significant increase in BMD [240], while in a 24-week study comparing isoflavone rich soy protein (80.4 mg aglycone isoflavones/ day) and isoflavone poor soy protein (4.4 mg aglycone isoflavones/ day) in perimenopausal women, both BMD and bone mineral content (BMC) were significantly higher with the diet high in isoflavones [238]. The lumbar spine seems to benefit the most from consumption of soy phytoestrogens. A 24-week long study with postmenopausal women consuming soy protein with 90mg isoflavones/day showed a significant increase in BMD of the lumbar spine, with no effect on the femoral neck or total body BMD [242].

Still in relation to isoflavones, studies have shown that ipriflavone, a synthetic isoflavone, effectively reduces bone loss in postmenopausal women [243]. This synthetic isoflavone undergoes intensive intestinal bacteria biotransformation to many metabolites, including daidzein [244]. It is believed that its action is exerted by interfering with bone resorption, and this is done by controlling preosteoclast recruitment, which decreases the number of osteoclasts and mononuclear cells. It is also believed that ipriflavone stimulates collagen synthesis, probably due to the activation of proline hydroxylase [245].

In addition to isoflavones, the effects of coumestans have also been investigated. An *in vitro* study using bone tissue in organ culture, reported that the exposure of 9-day old chick embryonic femurs to coumestrol inhibited bone resorption and stimulated bone mineralization. Both coumestrol and 17β estradiol inhibited the increased bone resorption induced by treatment with either parathyroid hormone, the hormonal form of vitamin D or prostaglandin E_2 [246].

The beneficial effects of soy isoflavones on bone tissue can result from either increased bone formation by osteoblasts or decreased bone resorption by osteoclasts. These mechanisms could increase bone mass, helping to prevent the development of osteoporosis. It has been suggested that osteoblasts and osteoclasts are target cells for the phytoestrogens. While estrogen receptors are present in relatively low numbers in osteoblasts, mammalian osteoclasts are not believed to express these receptors. Phytoestrogens may act directly on osteoblasts by genomic mechanisms, involving the activation or inhibition of the nuclear estrogen receptor, or, alternatively, may use non-genomic mechanisms, including the inhibition of tyrosine kinase [247], the inhibition of topoisomerase II [248] or the activation of a putative membrane-bound receptor for estrogenic molecules [249]. On the other hand, the inhibition of osteoclastic bone resorption may result from a direct action of phytoestrogens [250], presumably via non-genomic mechanisms, or from an indirect action mediated by inhibitory cytokines released by osteoblasts in response to the actions of phytoestrogens.

Despite the supportive studies regarding the benefits of a phytoestrogen rich diet on prevention of osteoporosis or bone fracture, several other studies have been unable to show the efficacy of phytoestrogen consumption on bone health. For instance, a study of 202 postmenopausal women found no benefit after 12 months of daily 99-mg isoflavone supplementation (52 mg of genistein, 41 mg of daidzein, and 6 mg of glycitein), which does not support the hypothesis that the use of a soy protein supplement containing isoflavones improves cognitive function, bone mineral density, or plasma lipids in healthy postmenopausal women when started at the age of 60 years or later [228,251].

Cardiovascular Disease

Several epidemiologic reports have linked the dietary intake of soy-based foods with the reduction of coronary heart disease [252]. In postmenopausal women, cardiovascular disease is the leading cause of death in many Western countries and it is known that estrogen can directly or indirectly affect the vascular system. In the first case, it acts through ERs located in vascular tissue; in the second case, it acts through altering the lipoprotein profile [253]. There are several clinical studies that have examined the effect that dietary phytoestrogens have on cardiovascular disease. Isoflavones or soy/soy protein and flaxseed have the ability to lower total cholesterol [242,254,255,256], low-density lipoprotein (LDL) cholesterol [242,255,256] and to raise high-density lipoprotein (HDL) [128,242]. Soy may affect the synthesis of cholesterol even in newborns, as male infants fed soybased formula had lower cholesterol fractional synthesis rates than infants fed breast milk or cow milk-based formula [257].

The improvement on LDL cholesterol levels in humans is directly related to the initial cholesterol concentration and depends on the amount of soy intake [258]. Improvement of HDL cholesterol is also directly proportional to the initial plasma HDL cholesterol concentration [259] and to gender [260]. A study assessing the effects of isoflavone-rich soy protein on blood lipids reported overall reductions in total cholesterol by 9.3%, triglycerides concentration by 10.5%, plasma LDL cholesterol concentration by 12.9% and increase in HDL cholesterol by 2% following an average daily consumption of 47 g of soy protein per day. The effects of dietary soy supplementation in lowering the lipid profile was found to be dose-related and more robust the highest the plasma cholesterol was at baseline [258]. Furthermore, isolated soy protein, rich in phytoestrogen, was found to enhance vascular reactivity in female monkeys with atherosclerosis, an effect similar to that observed with estrogen replacement therapy [18].

The effects of flaxseed on cardiovascular disease have also been linked to reductions in serum lipids. Several studies have reported a significant reduction in serum lipids, including total cholesterol and non-high density lipoprotein (non-HDL) cholesterol, apolipoprotein B and apolipoprotein A-I, as a result of flaxseed consumption [256,261].

Oxidized LDL particles are considered to be important in enhancing atherogenesis [262]. Due to the reported antioxidant property of phytoestrogens, several studies have evaluated the relative efficacy of isoflavones in inhibiting lipoprotein oxidation. For instance, in one study, equol and O-DMA were found to be more potent inhibitors of lipoprotein oxidation than genistein and daidzein, while, in another study, the relative antioxidant activity of the phytoestrogens was shown to be: genistein > daidzein = genistein = biochanin A = daidzin > formononetin.

Endothelium-derived nitric oxide (NO) is a potent vasodilator and mediates the effects of antihypertensive drugs such as nitroglycerin. Isoflavones have been shown to stimulate the activity of the endothelial NO synthase (NOS3), thus inducing vasodilation via NO [263,264]. Isoflavones have also been reported to have antithrombotic and antiatherogenic effects. For example, genistein and daidzein decrease monocyte chemoattractant protein-1 and collagen-induced platelet aggregation in a dose-dependent manner (264).

Several studies, however, showed no effect of isoflavones derived from soy or red clover on serum cholesterol levels [265] or plasma lipids [266]. Two studies that examined the dietary intake of lignans and isoflavones containing foods in postmenopausal women showed that the quartile with the highest intake had a more favorable waist to hip ratio, triglyceride levels, metabolic score [267] and aortic stiffness [268] than women with the lowest intake. No correlation was found between the intake of phytoestrogens and blood pressure, total, LDL and HDL cholesterol [267].

INHIBITION OF ANGIOGENESIS

Angiogenesis is the generation of new capillaries from pre-existing vessels. This process is absent in the healthy adult organism, being restricted to a few situations including wound healing and the formation of corpus luteum, endometrium and placenta [269]. In certain pathological conditions, angiogenesis increases and loses its self-limiting capacity [270], and this is particularly true for solid tumors [271]. Well-vascularized tumours expand both locally and by metastasis, while avascular tumours do not grow beyond a diameter of 1 – 2 mm [269].

The formation of new capillaries begins with a localized breakdown of the basement membrane of the parent vessel [272]. This is followed by a series of events that includes migration of endothelial cells and invasion of the surrounding matrix, elongation of the initial sprout, lumen formation and fusion of the capillary with the tip of another maturing sprout, forming a functional capillary loop [273]. Proteolytic degradation of the extracellular matrix by endothelial cells is controlled by angiogenic factors. For example, the basic fibroblast gowth factor (bFGF) induces the production of urokinase-type plasminogen activator [274] and its physiological inhibitor, plasminogen activator inhibitor-1 [272]. Studies using bovine microvascular endothelial cells have shown that genistein reduced both bFGF-stimulated and basal levels of both plasminogen activator and plasminogen activator inhibitor-1 activity in these cells [275]. This information suggests that phytoestrogens may contribute to the prevention of pathological neovascularization that is often seen during the development and progression of many diseases, including solid malignant tumors.

REPRODUCTIVE ACTIONS

Female Reproductive System

Phytoestrogens may alter ovarian cycles through ovarian, pituitary or hypothalamic actions. As mentioned earlier, genistein and coumestrol have a high affinity for ERβ [50], and this type of receptor is widely found in granulosa cells in the early stages of follicular development [276]. Therefore, phytoestrogens could inhibit estradiol action in the early stages of follicular development or augment it in the luteal phase. Clinical studies using vegetarian or Asian diets provided evidence for phytoestrogen influences on the human menstrual cycle although the results obtained were considerably varied. Both reductions and increases in estradiol secretion have been reported, along with lengthening or shortening of the menstrual cycle [116].

Indirect effects of phytoestrogens could also occur through the augmentation or suppression of estrogen negative feedback in the anterior pituitary or hypothalamus. Experimental studies have shown that genistein inhibited gonadotropin releasing hormone (GnRH)-stimulated LH release, suggesting an action at the level of the pituitary rather than the hypothalamus. On the other hand, genistein increased plasma prolactin, an estrogenic action caused by the inhibition of hypothalamic dopamine secretion [277,278]. Clinical studies have provided some evidence for the phytoestrogen regulation of gonadotropin release, but the results also were variable [279]. For example, in premenopausal women, reductions in midcycle LH and FSH were observed during a 1-month treatment with 0.8 mg/kg soy isoflavones [280], whereas in postmenopausal women, the same amount reduced LH levels over a 4-week period [281].

The effects of phytoestrogens on the uterine and vaginal epithelium have been reported in a variety of mammalian species. Epithelial cell proliferation in the uterus and vagina is a classical estrogen response, and isoflavones have been shown to induce uterine growth at oral doses of 100 – 800 mg/kg in mice [281], 10 - 40 mg/kg in rats [278,282]; and 100 mg/kg in cows [283]. However, a soy-based diet failed to induce uterine growth in ovariectomized rats [284] or rhesus macaques [285]. In addition, a soy-based diet with 10 mg/kg isoflavones did not induce maturation of the vaginal epithelium in ovariectomized rhesus monkeys [286] and also failed to alter vaginal cytology in ovariectomized rats [284].

Male Reproductive System

The effects of phytoestrogens on testicular function have been investigated in a few nonclinical trails with somewhat contradictory results. Micromolar concentrations of coumestrol (15μM) and genistein (7μM) reduced gonadotropin-induced testosterone production by cultured Leydig cells [287]. In pre-pubertal mice fed orally with 900 – 3600 mg/kg body weight for 6 weeks, genistin suppressed testicular weight and spermatogenesis, whereas oral doses of 300 - 1000 mg/kg genistein or genistin failed to alter testis weight over a 4-week period in adult male rats. Diets containing 7% soymeal increased testicular size when fed to pre-pubertal rats from weaning to 10 months of age and when fed to mice from fertilization to old age [116]. No significant differences in testosterone or testicular weight were found in rhesus monkeys treated with 1 mg/kg and 9 mg/kg soy isoflavones, respectively, for six months. In a similar dietary comparison, no effects of a higher isoflavone intake [288] on testicular weight were seen in juvenile rhesus monkey males ranging from 0.7 to 1.8 years of age [289].

Developmental Actions

The differentiation and development of the brain and reproductive tract is regulated by endogenous steroids, particularly in the early stages development [290,291,292], when their actions can permanently alter organ organization and subsequent cellular responsiveness. The action of non-steroidal estrogens during early development may be of some concern because they generally bind less well to the binding proteins that protect the developing offspring from maternal steroids and therefore have access to these tissues.

Although isoflavones concentrations in breast milk are generally lower than those found in adult plasma, concentrations in infant formula are high. Soy-based infant formulae contain concentrations at final dilution as high as 100 - 150 μM [293,294], and an infant nursed on human breast milk consumes no more than 4 μg/kg body weight per day. An infant on soy formula consumes as much as 6 - 9 mg/kg per day [294], a dose substantially higher than the 0.3 - 1.2 mg/kg dose received on a typical vegetarian or Asian diet [295]. Like other estrogenic substances, the isoflavones are effective differentiating agents in rodent models of development. For example, the prenatal administration of 20 - 100 mg/kg genistein to pregnant female rats showed reduced male anogenital distance and delayed vaginal opening in females [296,297]. Additionally, coumestrol and equol, at subcutaneous doses of 10 - 100 mg/kg, inhibited the development of uterine glands in female rats [298] and neonatal coumestrol induced precocious sexual maturation and premature anovulation following the treatment of lactating dams with oral doses of 16 mg/kg [299]. In mice, neonatal exposures to subcutaneous doses of 2 - 5 mg/kg coumestrol are associated with persistent vaginal cornification, haemorrhagic and polyovular follicles, uterine squamous metaplasia and cervicovaginal adenosis [300].

Early exposure to estrogen may also influence susceptibility to mammary cancer by enhancing or reducing the differentiation of terminal end-buds, the site of neoplastic transformation in the rodent mammary gland [301], thereby influencing susceptibility to carcinogenesis [302]. Similar effects have been observed with phytoestrogens. It has been reported that genistein enhances terminal duct differentiation and reduces the susceptibility to induced mammary cancer [303,304]. However, the high-dose treatments required for genistein action were also associated with precocious vaginal opening, atretic antral follicles and fewer corpora lutea [176].

Neurobehavioral Actions

In reference to brain, the majority of studies support the concept that isoflavones are neuroprotective [10]. The literature reports that soy phytoestrogens regulate choline acetyltransferase, nerve growth factor and brain-derived neurotrophic factor in brain areas such as the frontal cortex and hippocampus of female rats [305,306]. They have also been found to attenuate tau protein phosphorylations associated with Alzheimer's disease [307].

The presence of phytoestrogen in the brain was detected in adult Sprague – Dawley rats after intraperitoneal injection of genistein and daidzein [308]. Two well known hypothalamic brain regions are sexually dimorphic in rats. These two regions are the sexually dimorphic nucleus of the preoptic area (SDN-POA), which is normally two to five times larger in male than female rats, and the anteroventral periventricular (AVPV) nucleus, both under the influence of steroid hormones and/or estrogen-like molecules [309,310,311]. The SDN-POA is thought to regulate sexual behavior and gonadotropin levels [309,312], while the AVPV region regulates luteinizing hormone-releasing hormone via receptors for ovarian steroid hormones [311]. Studies have demonstrated that the perinatal administration of genistein significantly affected SDN-POA. In ovariectomized female rats, the injection of genistein significantly increased SDN-POA volumes [313,314], whereas in non-ovariectomized female rats, the administration of genistein resulted in a non-significant numerical decrease in SDN-POA volumes [296]. This suggests that isoflavones influence SDN-POA volumes that are dependent on the basal estrogen hormonal status of the organism [19,315].

The two major pathways of steroid metabolism in brain are the aromatization of androgens and the 5α-reduction of progesterone and androgens [316,317,318,319]. The aromatase cytochrome P450 (P450aro) is the enzyme that catalyzes the conversion of androgens to estrogens in specific brain regions [319,320,321], whereas, the 5α-reductase enzymes convert androgens, progestagens and glucocorticoids to their 5α-reduced metabolites throughout the brain [316,317,318]. The aromatase and 5α-reductase pathways are important mechanisms of seroid hormonal action in the central nervous system, especially in the hypoyhalamic region during perinatal and postnatal development. They also regulate neuroendocrine functions, reproductive endocrine physiology and sexual behavior [316,319,322,323].

The estrogen product of brain aromatase is known to stimulate neurite growth, migration and regulate neuronal plasticity in various brain regions; protect against neurodegenerative disease such as Alzheimer`s disease and stroke [324,325]; and influence learning, memory, stress, mood, reproductive endocrine function and sexual behavior [319,326,327]. The products of brain 5α-reductase are known to stimulate neurite growth [317,318,326,328]; promote neuroprotection in the spinal cord [330]; influence learning and memory [316,326,328] and inhibit epileptic seizures and alter mood, fatigue and locomotor activity [318,328,331]. Phytoestrogens are mild inhibitors of the P450aro and 5α-reductase enzymes in peripheral tissues. In the brain, experiments using animal models have shown that treatment with low phytoestrogen diets displayed varying levels of brain 5α-reductase, whereas administration of moderate (200 μg/g) or high (600 μg/g) phytoestrogen diets displayed stable brain 5α-reductase enzyme activities [332].

Calbindin-D28k (CALB) is a calcium-binding protein found in neurons thought to be responsible for sequestering/regulating intracellular calcium levels within homeostatic range and protecting neurons against apoptosis [333,334]. Decreased levels of CALCB may weaken calcium sequestration by the nerve cells leading to apoptosis as seen in neurodegenerative diseases such as Huntington's and Parkinson's diseases [333]. Several studies have shown that CALB levels were significantly decreased in the hypothalamus and amygdale of animals fed in a high phytoestrogen diet [334,335], suggesting a potential lack of neuroprotection.

Renal Disease

The mechanisms by which phytoestrogens may exert their renal protective effects are not completely elucidated. However, the available data suggests that diets rich in isoflavones and lignans, namely soy protein and flaxseed, can slow the development and progression of renal disease as observed in various types of chronic renal disease in both animals and humans. Several mechanisms have been proposed to explain the renal protective effects of isoflavones and lignans They include inhibition of cell growth and proliferation through ER-mediated mechanisms, especially ERβ, or through non-ER-mediated pathways; inhibition of tyrosine kinases and DNA topoisomerases I and II; inhibition of angiogenesis; modulation of TGF-β1 and other growth factors; inhibition of cytokine-mediated activation of transcription factors; and inhibition of platelet-activating factor (PAF) [168]. In humans, the consumption of a vegetarian diet with 52% soybean protein during eight weeks reduced proteinuria, serum lipids and serum lipoproteins in cases of chronic glomerular disease with nephritic syndrome [336], and reduced proteinuria and hyperlidemia in cases of chronic glomerulonephritis with nephritic-range proteinuria [337]. In cases of lupus nephritis, decreased serum lipids, proteinuria, increased creatinine clearance and inhibition of PAF were observed after a 12-week ingestion of 15, 30 and 45 g/d of flaxseed sequentially at 4-week intervals [338].

Anti-Viral Activity

In addition to all beneficial effects already mentioned, phytoestrogens have also been found to exert anti-viral properties. Viruses enter cells in order to carry out replication and they do so by a variety of mechanisms, such as clathrin-mediated endocytosis and caveolae-mediated endocytosis, which have been shown to be influenced by phytoestrogens [339,340]. In caveolae-mediated endocytosis, some viruses, before internalization, firstly attach to major histocompatibility complex class I molecules on the cell surface, move laterally along the plasma membrane and eventually become trapped in a caveola, which is an invagination of the plasma membrane lined with a protein coat of caveolin-1. This is followed by activation of tyrosine kinases and since phytoestrogens are known to be tyrosine kinase inhibitors, it is believed that they inhibit the phosphorylation events important for viral entry and infection [49,247]. For example, in kidney fibroblast cells treated with genistein, simian virus 40 (SV-40) (associated with pleural mesothelioma and osteosarcome) were unable to enter the cells due to inhibition of virus-induced signals that are necessary for its entry. Evidence suggests that genistein may block tyrosine phosphorylation of caveolin-1 resulting in inhibition of SV-40 entry into the cells [341,342]. Following internalization, the virus is transported from the caveosome to the endoplasmic reticulum. Genistein have also been found to block the downstream signals that enable the SV-40 virus to reach the endoplasmic reticulum [343] (Figure 15).

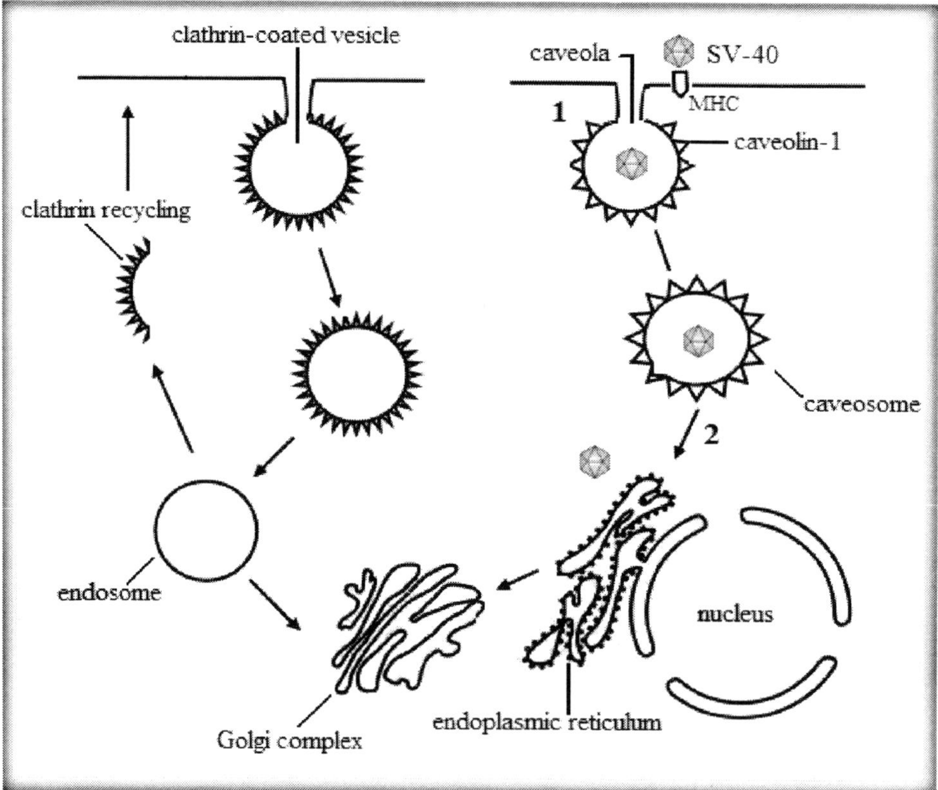

Figure 15. Schematic presentation of clathrin-mediated endocytosis and caveola-mediated endocytosis. It is suggested that genistein may block tyrosine phosphorylation of caveolin -1 resulting in inhibition of SV-40 entry into the cells (**1**) [342]. Following internalization, the virus is transported from the caveosome to the endoplasmic reticulum. Genistein have also been found to block the downstream signals that enable the SV-40 virus to reach the endoplasmic reticulum (**2**) [343]. (Based on reference 9).

In another study, 50μM genistein was capable of suppressing replication of encephalomyocarditis virus in L929 cells to 0.5% of control level by a mechanism involving inhibition of viral protein synthesis [344]. When dengue virus infects cells, a series of events occur leading to alteration in cellular permeability with associated actin cytoskeleton rearrangements. Human dermal microvascular endothelial cells treated with genistein showed inhibition of phosphorylation of proteins involved with actin reorganization causing a reduction in stress fibers and focal adhesions during dengue virus infection [345]. Biochanin A has also been shown to exert anti-viral activity. In HSB-2 cells exposed to human herpesvirus 6, biochanin A blocked infection by inhibiting the expression of early and late viral antigens and syncytium formation [346].

Few animal studies have provided evidence to corroborate the *in vitro* findings. In pigs, it has been shown that a soy genistein enriched diet reduced replication of porcine reproductive and respiratory syndrome virus when virally challenged [347]. Although genistein is the best characterized phytoestrogen in regard to its anti-viral activity, other types of phytoestrogens are expected to exert similar function and this further supports the health benefits of a phytoestrogen-rich diet in humans.

PHYTOESTROGEN TOXICITY

Although most phytoestrogens exert pleiotropic effects involving kinase inhibition, cell cycle regulation and antioxidative properties that are likely to contribute to the beneficial effects, estrogenic and/or anti-estrogenic activities may reduce but also may stimulate estrogen-dependent tumor growth depending on dose and timing of exposure. Other adverse effects have been reported in relation to the action of phytoestrogens. For example, in mice neonates exposed to genistein at dose levels comparable to serum levels of 1 – 5 μM in human infants, a variety of adverse reproductive effects were noted later in life, including altered mammary gland morphology, abnormal estrous cycle, altered ovarian function and infertility [348]. *In vitro* cell culture studies have reported genistein to be clastogenic (chromosome breaking), DNA damaging and even mutagenic, in contrast to *in vivo* studies that generally have shown negative genotoxicity results [349]. In Chinese hamster V79 cells, 50 – 150 μM genistein was found to induce micronucleus formation, indicative of chromosome breakage [350].

Another situation which deserves attention is the feeding of neonates with soy-products. Soy infant formula contains high levels of the isoflavones genistein and daidzein that are well absorbed by infants and circulate at concentrations that are greater than levels which have been shown to produce physiological effects in adult women consuming a high soy diet, generating considerable concern regarding their long-term health effects, particularly on endocrinological and reproductive outcomes. The existing data indicate no adverse effect on human growth, development or reproduction [351], however there is very little research on the effects of consumption of soy phytoestrogens by human neonates and this requires a cautious approach in situations where there are potential developmental effects from the consumption of pharmacologically active compounds in infancy and childhood [352].

Conclusion

In vitro and *in vivo* studies indicate that phytoestrogens exhibit a broad spectrum of biological activities and resemble other estrogens in their ability to regulate cell proliferation and differentiation. Phytoestrogens also exhibit anti-proliferative and anti-estrogenic actions that are thought to provide protection against carcinogenesis. Initially, it was thought that phytoestrogen actions were predominantly hormonal, but it is now clear that they display a diversity of properties that enable them to exert multiple effects in the organism, and this is consistent with steroid action in general. Differences in the amounts and distribution of α and β estrogen receptors reflect different responses in the cell since phytoestrogens have preferential actions on either receptor. The discrepancies observed in the many studies available are probably related to expected limitations of dietary assessments. This is particularly true in that they may fail to consider variations in the bioavailability of phytoestrogens due to, for example, intestinal microflora, sex, genetics, diet and antibiotic use. Human responses to phytoestrogens were shown to be highly variable in both quality and incidence of response as well as in the nature of the hormonal response as both estrogenic and anti-estrogenic responses have been observed. Phytoestrogens preparations, as pharmaceuticals and herbal medicines or as food supplements, require quality standards regarding manufacturing, and special attention should be given to isoflavone content. The growing availability and popularity of purified phytoestrogen makes it especially important to investigate even further the possibility of their beneficial or adverse effects.

References

[1] Mazur, W. Phytoestrogen content in foods. Bail. *Clin. Endocrinol. Metab.*, 1998 12, 729-742.

[2] Steinmetz, KA; Potter, JD. Vegetables, fruit and cancer. I, Epidemiology. *Cancer Causes and Control*, 1991a 2, 325-357.

[3] Steinmetz, KA; Potter, JD. Vegetables, fruit and cancer. II, Mechanisms. *Cancer Causes and Control,* 1991b 2, 427-442.

[4] Dragsted, LO; Strube, M; Larsen, JC. Cancer-protective factors in fruits and vegetables, biochemical and biological background. *Pharmacol. Toxicol.,* 1993 suppl. 72, 116-135.

[5] Setchell, KDR. Phytoestrogens, biochemistry, physiology and implications for human health of soy isoflavones. *Am. J. Clin. Nutr.*, 1998 129, 1333S–1346S.

[6] Adlercreutz, H; Mazur, W. Phyto-estrogens and western diseases. *Ann. Med.*, 1997 29, 95–120.

[7] Barnes, S. Evolution of the health benefits of soy isoflavones. *Proc. Soc. Exp. Biol. Med.*, 1998 217, 386–392.

[8] Kurzer, MS; Xu, X. Dietary phytoestrogens. *Ann. Rev. Nutr.*, 1997 17, 353–381.

[9] Martin, JHJ; Crotty, S; Warren, P; Nelson, PN. Does an apple a day keep the doctor away because a pytoestrogen a day keeps the virus at bay? A review of the anti-viral properties of phytoestrogens. *Phytochemistry*, 2007 68, 266-274.

[10] Lephart, ED; Setchell, KDR; Lund, TD. Phytoestrogens: hormonal action and brain plasticity. *Brain Res. Bull.*, 2005 65, 193-198.

[11] Adlercreutz, H. Phyto-oestrogens and cancer. Lancet Oncology, 2002 3, 364–373.
[12] Whitten, PL; Patisaul, HB. Cross-species and interassay comparisons of phytoestrogen action. *Environ. Health Perspect.*, 2001 suppl. 109, 5–20.
[13] Glazier, MG; Bowman, MA. A review of the evidence for the use of phytoestrogens as a replacement for traditional estrogen replacement therapy. *Arch. Intern. Med.*, 2001 161, 1161– 1172.
[14] Knight, DC; Eden, JA. A review of the clinical effects of phytoestrogens. *Obstet. Gynecol.*, 1996 87, 897– 904.
[15] Mazur, W; Adlercreutz, H. Overview of naturally occurring endocrineactive substances in the human diet in relation to human health. *Nutrition*, 2000 16, 654–658.
[16] Murkies, AL; Wilcox, G; Davis, SR. Phytoestrogens—review. *J. Clin. Endocrinol. Metab.*, 1998 83, 297–303.
[17] Setchell, KDR. A. Cassidy, Dietary isoflavones—biological effects and relevance to human health. *J. Nutr.*, 1999 129, 758S– 767S.
[18] Albertazzi, P; Purdie, D. The nature and utility of the phytoestrogens, a review of the evidence. *Maturitas*, 2002 42, 173–185.
[19] Lephart, ED; West, TW; Weber, KS; Rhees, RW; Setchell, KD; Adlercreutz, H; Lund, TD. Neurobehavioral effects of dietary soy phytoestrogens. *Neurotox. Terat.*, 2002 24, 5–16.
[20] Kalimi, M; Dave, JR. Mechanisms of action of hormones associated with reproduction. In: Witorsch, RJ, editor. *Reproductive Biology*. New York: Raven Press, 1995, 61-72.
[21] Hess, RA; Bunick, D; Lee, K-H; Bahr, J; Taylor, JA; Korach, KS; Lubahn, DB. A role for oestrogens in the male reproductive system. *Nature*, 1997 390**,** 509–512.
[22] Sharpe, RM. Do males rely on female hormones? *Nature*, 1997 390**,** 447–448.
[23] Farhat, MY; Lavigne, MC; Ramwell, PW. The vascular protective effects of estrogen. *FASEB J.*, 1996 10, 615–624.
[24] Turner, RT; Riggs, BL; Spelsberg, TC. Skeletal effects of estrogen. *Endocrine Rev.*, 1994; 15, 275–296.
[25] Kuiper, GGJM; Shughrue, PJ; Merchenthaler, I; Gustafsson, J-Å. The estrogen receptor β subtype: a novel mediator of estrogen action in neuroendocrine systems. *Front. Neuroendocrinol.*, 1998 19, 253-286.
[26] Joels, M. Steroid hormones and excitability in the mammalian brain. *Front. Neuroendocrinol.,* 1997 18, 2–48.
[27] Wickelgren, I. Estrogen stakes claim to cognition. *Science*, 1997 276**,** 675–678.
[28] Kincade, PW; Medina, KL; Smithson, G. Sex hormones as negative regulators of lymphopoiesis. *Immunological Rev.*, 1994 137**,** 119–134.
[29] Herbert, DC; Supakar, PC; Roy, AK. Male reproduction. In: Witorsch, RJ, editor. *Reproductive Biology*. New York: Raven Press, 1995, 3-21.
[30] Hall, JM; Couse, JF; Korach, KS. The multifaceted mechanisms of estradiol and estrogen receptor signaling. *J. Biol. Chem.*, 2001 276, 36869–36872.
[31] Evans, R. The steroid and thyroid hormone receptor superfamily. *Science*, 1988 240, 889-895.
[32] Carson-Jurica, M; Schrader, W; O`Malley, B. Steroid receptor family: structure and function. *Endocrine Rev.*, 1990 11, 201-220.
[33] Wahli, W; Martinez, E. Superfamily of steroid nuclear receptor: positive abd negative regulators of gene expression. *FASEB J.*, 1991 5, 2243-2249.

[34] Fuller, P. The steroid receptor superfamily: mechanism of diversity. *FASEB J.*, 1991 5, 3092-3099.
[35] Cornwell, T; Cohick, W; Raskin, I. Dietary phytoestrogens and health. *Phytochemistry*, 2004 65, 995-1016.
[36] Anderson, JJB; Garner, SC. Phytoestrogens and bone. *Bail. Clin. Endocrinol. Metab.*, 1998 12, 543-557.
[37] Xu, YJ; Traystman, RD; Hurn, P; Wang, M. Neuritelocalized estrogen receptor—a mediates rapid signaling by estrogen. *J. Neurosci. Res.*, 2003 74, 1–11.
[38] Chen, D-B; Bird, IM; Zheng, J; Magness, RR. Membrane estrogen receptor-dependent extracellular signal-regulated kinase pathway mediates acute activation of endothelial nitric oxide synthase by estrogen in uterine artery endothelial cells. *Endocrinology*, 2004 145, 113–125.
[39] Brann, DW; Hendry, LB; Mahesh, VB. Emerging diversities in the mechanism of action of steroid hormones. *J. Steroid. Biochem. Mol. Biol.*, 1995 52, 113–133.
[40] Revelli, A; Massobrio, M; Tesarik, J. Non-genomic actions of steroid hormones in reproductive tissues. *Endocrine Rev.*, 1998 19, 3–17.
[41] Wehling, M. Specific, nongenomic actions of steroid hormones. *Annu. Rev. Physiol.*, 1997 59, 365–393.
[42] Beato, M; Herrlich P; Schutz, G. Steroid hormone receptors, Many actors in search of a plot. *Cell*, 1995 83, 851–857.
[43] Zilliacus, J; Carlstedt-Duke, J; Gustafsson, J-Å; Wright, APH. Evolution of distinct DNA binding specificities within the nuclear receptor family of transcription factors. *Proc. Natl. Acad. Sci. USA*, 1994 91, 4175–4179.
[44] Horwitz, KB; Jackson, TA; Bain, DL; Richer, JK; Takimoto, GS; Tung, L. Nuclear receptor coactivators and corepressors. *Mol. Endocrinol.*, 1996 10, 1167–1177.
[45] Tsai, M-J; O'Malley, BW. Molecular mechanisms of action of steroid/thyroid receptor superfamily members. *Annu. Rev. Biochem.*, 1994 63, 451–486.
[46] Nilsson, S; Gustafsson, JA. Biological role of estrogen and estrogen receptors. Crit. Rev. Biochem. *Mol. Biol.*, 2002 37, 1–28.
[47] Macgregor, JI; Jordan, VC. Basic guide to the mechanisms of antiestrogen action. *Pharmacol. Rev.*, 1998 50, 151–196.
[48] Couse, JF; Hewitt, SC; Bunch, DO; Sar, M; Walker, VR; Davis, BJ; Korach, KS. Postnatal sex reversal of the ovaries in mice lacking estrogen receptors α and β. *Science*. 1999 286, 2328–2331.
[49] Liu, MM; Albanese, C; Anderson, CM; Hilty, K; Webb, P; Uht, RM; Price, RH; Pestell, RG; Kushner, PJ. Opposing action of estrogen receptors α and β on cyclin D1 gene expression. *J. Biol. Chem.*, 2002 277, 24353–24360.
[50] Kuiper, GGJM; Carlsson, B; Grandien, K; Enmark, E; Haggblad, J;., Nilsson, S; Gustafsson, JA. Comparison of the ligand binding specificity and transcript tissue distribution of estrogen receptors α and β. *Endocrinology*, 1997 138, 863–870.
[51] Turner, RT; Riggs, BL; Spelsberg, TC. Skeletal effects of estrogen. *Endocrine Rev.*, 1994 15, 275–296.
[52] Onoe, Y; Miyaura, C; Ohta, H; Nozawa, S; Suda, T. Expression of estrogen receptor β in rat bone. *Endocrinology*, 1997 138, 4509–4512.

[53] Kuiper, GG; Lemmen, JG; Carlsson, B; Corton, JC; Safe, SH; van der Saag, PT; van der Burg, B; Gustafsson, JA. Interaction of estrogenic chemicals and phytoestrogens with estrogen receptor beta. *Neuroendocrinology*, 1998 139, 4252–4263.
[54] Setchell, KDR. Soy isoflavones-benefits and risks from nature's selective estrogen receptor modulators (SERMs). *J. Am. Coll. Nutr.*, 2001 20, 354S–383S.
[55] Halbreich, U; Kahn, LS. Selective oestrogen receptor modulators–current and future brain and behaviour applications. *Expert. Opin. Pharmacother.*, 2000 1, 1385–1398.
[56] Raffaeli, B; Hoikkala, E; Leppälä, E; Wähälä, K. Enterolignans. *J. Chromatr.*, 2002 777, 29-43.
[57] Lotke, PS. Phytoestrogens: a potential role in hormone replacement therapy. *Prim. Care Update Ob/Gyns.*, 1998 5, 290-295.
[58] Adams, NRJ. A changed responsiveness to oestrogens in ewes with clover disease. *Reprod. Fertil.*, 1981 suppl. 30, 223-230.
[59] Leopold, AS; Erwin, M; Oh, J; Browning, B. Phytoestrogens, adverse effects on reproduction in California quail. *Science*, 1976 191, 98–99.
[60] Setchell, KD; Gosselin, SJ; Welsh, MB; Johnston, JO; Balistreri, WF; Kramer, LW; Dresser, BL; Tarr, M. Dietary estrogens-a probable cause of infertility and liver disease in captive cheetahs. *Gastroenterology*, 1987a 93, 225–233.
[61] Setchell, KD; Gosselin, SJ; Welsh, MB; Johnston, JO; Balistreri, WF; Kramer, LW; Dresser, BL; Tarr, MJ. Dietary estrogens – a probable cause of infertility and liver disease in captive cheetahs. *Gastroenterology*, 1987b 93, 225–233.
[62] Dixon, RA; Ferreira, D. Genistein. *Phytochemistry*, 2002 60, 205-211.
[63] Reinli, K; Block, G. Phytoestrogen content of foods—a compendium of literature values. *Nutr. Cancer*, 1996 26, 123–148.
[64] Boker, LK; Van der Schouw, YT; De Kleijn, MJ; Jacques, PF; Grobbee, DE; Peeters, PH. Intake of dietary phytoestrogens by Dutch women. J. Nutr., 2002 132, 1319–1328.
[65] Wang, HJ; Murphy, PA. Isoflavone content in commercial soybean foods. *J. Agric. Food Chem.*, 1994 42, 1666–1673.
[66] Wang, HJ; Murphy, PA. Isoflavone composition of American and Japanese soybeans in Iowa, effects of variety, crop year, and location. *J. Agric. Food Chem.*, 1994 42, 1674–1677.
[67] Stevens, JF; Page, JE. Xanthumol and related prenylflavonoids from hops and beer: to your good health! *Phytochemistry*, 2004 65, 1317-1330.
[68] Birt, DF; Hendrich, S; Wang, W. Dietary agents in cancer prevention: flavonoids and isoflavonoids. *Pharmacol. Ther.*, 2001 90, 157-177.
[69] Dixon, RA; Ferreira, D. Genistein. *Phytochemistry*, 2002 60, 205– 211.
[70] Hodgson, JM; Croft, KD; Puddey, IB et al. Soybean isoflavonoids and their metabolic products inhibit in vitro lipoprotein oxidation in serum. *J. Nutr. Biochem.*, 1996 7, 664-669.
[71] Zhang, Y; Wang, G-J; Song, TT; Murphy, PA; Hendrich, S. Urinary disposition of the soybean isoflavones, daidzein, genistein and glycitein differs among humans with moderate fecal isoflavone degradation activity. *J. Nutr.*, 1999 129, 957-962.
[72] Adlercreutz, H; Fotsis, T; Kurzer, MS; Wahala, K; Makela, T; Brunow, G; Hase, T. Isotope dilution gas chromatographic- mass spectrometric method for the determination

of unconjugated lignans and isoflavonoids in human feces, with preliminary results in omnivorous and vegetarian women. *Anal. Biochem.,* 1995 225, 101-108,

[73] Adlercreutz, H; Fotsis ,T; Lampe, J et al. Quantitative determination of lignans and isoflavonoids in plasma of omnivorous and vegetarian women by isotope dilution gas chromatography-mass spectrometry. Scandinavian *J. Clin. Lab. Invest.*, 1993 215, 5–18.

[74] Karr, SC; Lampe, JW; Hutchins, AM; Slavil, JL. Urinary isoflavonoid excretion in humans is dose dependent at low to moderate levels of soy-protein consumption. *Am. J. Clin. Nutr.*, 1997 66, 46–51.

[75] Setchel, KDR; Borriello, SP; Hulme, P; Kirk, DN; Axelson, M. Non-steroidal estrogens of dietary origin, a possible role in gut hormone metabolism. *Am. J. Clin. Nutr.*, 1984 40, 569–578.

[76] Morton, MS; Wilcox, G; Wahlgvist, ML; Griffitts, K. Determination of lignans and isoflavonoids in human female plasma following dietary supplementation. *J. Endocrinol.*, 1994 142, 251-259.

[77] Xu, X; Wang, H-J; Murphy, PA; Cook, L; Hendrich, S. Daidzein is a more bioavailable soymilk isoflavone than is genistein in adult women. *J. Nutr.*, 1994 124, 825-832,

[78] King, RA; Bursill, DB. Plasma and urinary kinetics of the isoflavones daidzein and genistein after a single soy meal in humans. *Am. J. Clin. Nutr.*, 1998 67, 867-872.

[79] Watanabe, S; Yamaguchi, M; Sobue, T; Miura, T; Arai, Y; Mazur, W, Wahala, K; Adlercreutz, H. Pharmacokinetics of soybean isoflavones in plasma, urine, and feces of men after ingestion of 60 g baked soybean powder (kinako). *J. Nutr.*, 1998 128, 1710-1715.

[80] Kudou, S; Fleury, Y; Welti, D et al. Malonyl isoflavone glycosides in soybean seeds (Glycine max MERRILL). *Agric. Biol. Chem.*, 1991 55, 2227-2233.

[81] Barnes, S; Kirk, M; Coward L. Isoflavones and their conjugates in soy foods, extraction conditions and analysis by HPLC-mass spectrometry. *J. Agric. Food Chem.*, 1994 42, 2466-2474.

[82] Farmakalidis, E; Murphy, PA. Isolation of 6"-O-acetyldaidzein and 6"-O-acetylgenistein from toasted defatted soy flakes. *J. Agric. Food Chem.,* 1985 33, 385-389.

[83] Coward, L; Barnes, NC; Setchell, KDR; Barnes, S. The antitumor isoflavones, genistein and daidzein, in soybean foods of American and Asian diets. *J. Agric. Food Chem.,* 1993 41, 1961-1967.

[84] King, RA; Broadbent, JL; Head, RJ. Absorption and excretion of the soy isoflavone genisteinin rats. *J. Nutr.,* 1996 126, 176-182.

[85] Sfakianos, J; Coward, L; Kirk, M; Barnes, S. Intestinal uptake and biliary excretion of the isoflavone genistein in the rat. *J. Nutr.*, 1997 127, 1260-1268.

[86] Farmakalidis, E; Hathcock, JN; Murphy, PA. Oestrogenic potency of genistin and daidzin in mice. *Food Chem. Toxicol.,* 1985 23, 741–745.

[87] Markiewicz, L; Garey, J; Adlercreutz, H; Gurpide, E. In vitro bioassays of non-steroidal phytoestrogens. J. Steroid Biochem. *Mol. Biol.*, 1993 45, 399–405.

[88] Molteni, A; Brizio-Molteni, L; Persky, V. In vitro hormonal effects of soybean isoflavones. *J. Nutr.*, 1995 125, 751S–756S.

[89] Nagel, SE; vonSaal, FF; Welshons, WV. The effective free fraction of estradiol and xenoestrogens in human serum measured by whole cell uptake assays; physiology of delivery modifies estrogenic activity. *Proc. Soc. Exp. Biol. Med.*, 1998 217, 300–309.

[90] Hol, T; Cox, MB; Bryant, HU; Draper, MW. Selective estrogen receptor modulators and postmenopausal women's health. *J. Women's Health,* 1997 6, 523– 531.

[91] Liu, J; Burdette, JE; Xu, H; Gu, C; van Breemen, RB; Bhat, KP; Booth, N; Constantinou, AI; Pezzuto, JM; Fong, HH; Farnsworth, NR; Bolton, JL. Evaluation of estrogenic activity of plant extracts for the potential treatment of menopausal symptoms. *J. Agric. Food Chem.*, 2001 49, 2472–2479.

[92] Davis, JN; Kucuk, O; Sarkar, FH. Expression of prostatespecific antigen is transcriptionally regulated by genistein in prostatecancer cells. *Mol. Carcinog.*, 2002 34, 91–101.

[93] Wang, D; Gutkowska, J; Marcinkiewicz, M; Rachelska, G; Jankowski, M. Genistein supplementation stimulates the oxytocin system in the aorta of ovariectomized rats. *Cardiovasc. Res.*, 2003. 57, 186–194.

[94] Zhou, S; Turgeman, G; Harris Stephen, E; Leitman Dale, C., Komm Barry, S; Bodine Peter, V.N; Gazit, D. Estrogens activate bone morphogenetic protein-2 gene transcription in mouse mesenchymal stem cells. *Mol. Endocrinol.,* 2003 17, 56–66.

[95] Murrill, W; Brown, N; Zhang, J; Manzolillo, P; Barnes, S; Lamartiniere, CA. Prepubertal genistein exposure suppresses mammary cancer and enhances gland differentiation in rats. *Carcinogenesis*, 1996 17, 1451–1457.

[96] Fritz, WA; Wang, J; Lamartiniere, CA. Dietary genistein, perinatal mammary cancer prevention, bioavailability and toxicity testing in the rat. *Carcinogenesis*, 1998 19, 2151–2158.

[97] Fritz, WA; Eltoum, I-E; Cotroneo, MS; Lamartiniere, CA. Genistein alters growth but is not toxic to the rat prostate. *J. Nutr.,* 2002 132, 3007–3011.

[98] Yanagihara, K; Ito, A; Toge, T; Numoto, M. Antiproliferative effects of isoflavones on human cancer cell lines established from the gastrointestinal tract. *Cancer. Res.*, 1993 53, 5815–5821.

[99] Cotroneo, MS; Wang, W; Eltoum, IE; Lamartiniere, CA. Sex steroid receptor regulation by genistein in the prepubertal rat uterus. *Mol. Cell. Endocrinol.,* 2001 173, 135–145.

[100] Strauss, L; Makela, S; Joshi, S; Huhtaniemi, I; Santti, R; Genistein exerts estrogen-like effects in male mouse reproductive tract. *Mol. Cell. Endocrinol.,* 1998 144, 83–93.

[101] Fritz, WA; Wang, J; Eltoum, I-E; Lamartiniere, CA. Dietary genistein down-regulates androgen and estrogen receptor expression in the rat prostate. *Mol. Cell. Endocrinol.,* 2002 186, 89-99.

[102] Thompson, LU; Robb, P; Serraino, M; Cheung F. Mammalian lignan production from various foods. *Nutr.Cancer*, 1991 16, 43–52.

[103] Thompson, LU; Rickard, SE; Cheung, F et al. Variability in anticancer lignan levels in flaxseed. *Nutr. Cancer*, 1997 27, 26–30.

[104] Heinonen, S; Nurmi, T; Liukkonen, K; Poutanen, K; Wahala, K; Deyama T et al. In vitro metabolism of plant lignans, new precursors of mammalian lignans enterolactone and enterodiol. *J. Agric. Food Chem.*, 2001 49, 3178- 3186.

[105] Adlercreutz, H; Wang, GJ; Lapcik, O; Hampl, R; Wahala, K; Makela, T; Lusa, K; Talme, M; Mikola, H. Time-resolved fluoroimmunoassay for plasma enterolactone. *Anal. Biochem.*, 1998 265, 208–215.

[106] Dean, B; Chang, S; Doss, GA; King, C; Thomas, PE. Glucuronidation, oxidative metabolism, and bioactivation of enterolactone in rhesus monkeys. *Arch. Biochem. Biophys.*, 2004 429, 244-251.

[107] Setchell, KDR; Adlercreutz H. Mammalian lignans and phyto-estrogens. Recent studies on their formation, metabolism and biological role in health and disease. In: Rowland, IR, editor. *Role of the Gut Flora in Toxicity and Cancer.* London: Academic Press, 1988.

[108] Lampe, JW; Martini, MC; Kurzer, MS et al. Urinary lignan and isoflavonoid excretion in premenopausal women consuming flaxseed powder. *Am. J. Clin. Nutr.*, 1994 60, 122–128.

[109] Nesbitt, PD; Lam, Y; Thompson, LU. Human metabolism of mammalian lignan precursors in raw and processed flaxseed. *Am. J. Clin. Nutr,* 1999 69, 549–555.

[110] Duncun, AM; Phipps, WR; Kurzer, MS. Phyto-oestrogens. Best Pract. Res. *Clin. Endocrinol. Metab.*, 2003 17, 273-271.

[111] Horner, NK; Kristal, AR; Prunty, J et al. Dietary determinants of plasma enterolactone. Cancer Epidemiol. *Biomarkers Prev.,* 2002 11, 121–126.

[112] Murphy, PA; Song, T., Buseman, G; Barua, K; Beecher, GR; Trainer, D; Holden, J. Isoflavones in retail and institutional soy foods. *J. Agric. Food Chem.*, 1999 47, 2697–2704.

[113] Wang, GS; Kuan, S; Francis, OJ; Ware, GM; Carman, AS. A simplified HPLC method for the determination of phytoestrogens in soybean and its processed products. *J. Agric. Food Chem.*, 1990 38, 185–190.

[114] Franke, AA; Custer, LJ; Cerna, CM; Narala, KK. Quantitation of phytoestrogens in legumes by HPLC. *J. Agric. Food Chem.*, 1994 42, 1905–1913.

[115] de Kleijn, MJ; van der Schouw, YT; Wilson, PW; Adlercreutz, H; Mazur, W; Grobbee, DE; Jacques, PF. Intake of dietary phytoestrogens is low in postmenopausal women in the United States. The Framingham study. *J. Nutr.*, 2001 131, 1826–1832.

[116] Whitten, PL; Naftolin, F. reproductive actions of phytoestrogens. *Bail. Clin. Endocrinol. Metab.*, 1998 12, 667-690.

[117] Milligan, S; Kalita, J; Pocock, V; Heyerick, A; De Cooman, L; Rong, H; De Keukeleire, D. *Oestrogenic activity of the hop phyto-oestrogen, 8-prenylnaringenin.* Reproduction Bristol UK, 2002 123, 235–242.

[118] Vinson, JA; Mandarano, M; Hirst, M; Trevithick, JR; Bose, P. Phenol antioxidant quantity and quality in foods, beers and the effect of two types of beer on an animal model of atherosclerosis. *J. Agric. Food Chem.*, 2003 51, 5528–5533.

[119] Stevens, JF; Miranda, CL; Frei, B; Buhler, DR. Inhibition of peroxynitrite-mediated LDL oxidation by prenylated flavonoids, the alpha, beta-unsaturated keto functionality of 20-hydroxychalcones as a novel antioxidant pharmacophore. *Chem. Res. Toxicol.,* 2003 16, 1277–1286.

[120] Schaefer, O; Humpel, M; Fritzemeier, KH; Bohlmann, R., Schleuning, WD. 8-Prenylnaringenin is a potent ERalpha selective phytoestrogen present in hops and beer. *J. Steroid Biochem. Mol. Biol.*, 2003 84, 359–360.

[121] Schmitt, E; Lehmann, L; Metzler, M; Stopper, H. Hormonal and genotoxic activity of resveratrol. Toxicol. Let., 2002 136, 133-142.

[122] Gehm, BD; McAndrews, JM; Chien, PY; Jameson, JL. *Resveratrol, a polyphenolic compound found in grapes and wine, is an agonist for the estrogen receptor. PNAS,* 1997 94, 14138–14143.

[123] Frankel, EN; Waterhouse, AL; Teissedre, PL. Principal phenolic phytochemicals in selected California wines and their antioxidant activity in inhibiting oxidation of human low-density lipoproteins. *J. Agric. Food Chem.*, 1995 43, 890–894.

[124] Creasy, LL; Creasy, MT. Grape chemistry and the significance of resveratrol, An Overview. *Pharm. Biol.*, 1998 36, 8–14.

[125] Fremont, L. Biological effects of resveratrol. Life Sci., 2000 66, 663– 673,

[126] Lamikanra, O; Grimm, CC; Rodin, JB; Inyang, ID. Hydroxylated stilbenes in selected american wines. *J. Agric. Food Chem.*, 1996 44, 1111–1115.

[127] Sobolev, VS; Cole, RJ. trans-Resveratrol content in commercial peanuts and peanut products. *J. Agric. Food Chem.*, 1999 47, 1435–1439.

[128] Sanders, TAB; Dean, TS; Grainger, D; Miller, GJ; Wiseman, H. Moderate intakes of intact soy protein rich in isoflavones compared with ethanol-extracted soy protein increase HDL but do not influence transforming growth factor beta(1) concentrations and hemostatic risk factors for coronary heart disease in healthy subjects. *Am. J. Clin. Nutr.*, 2002 76, 373–377.

[129] Vastano, BC; Chen, Y; Zhu, N; Ho, C-T; Zhou, Z; Rosen, RT. Isolation and identification of stilbenes in two varieties of *Polygonum cuspidatum. J. Agric. Food Chem.*, 2000 48, 253–256.

[130] Powell, RG; TePaske, MR; Plattner, RD; White, JF; Clement, SL. Isolation of resveratrol from *Festuca versuta* and evidence for the widespread occurrence of this stilbene in the poaceae. *Phytochemistry*, 1994 35, 335–338.

[131] Mannila, E; Talvitie, A. Stilbenes from Picea abies bark. *Phytochemistry*, 1992 31, 3288–3289.

[132] Klinge, CM; Risinger, KE; Watts, MB; Beck, V; Eder, R; Jungbauer, A. Estrogenic activity in white and red wine extracts. *J. Agric. Food Chem.*, 2003 51, 1850–1857.

[133] Bowers, JL; Tyulmenkov, VV; Jernigan, S.C; Klinge, CM. Resveratrol acts as a mixed agonist/antagonist for estrogen receptors alpha and beta. *Endocrinology*, 2000 141, 3657–3667.

[134] Brownson, DM; Azios, NG; Fuqua, BK; Dharmawardhane, SF; Mabry, TJ. Flavonoid effects relevant to cancer. *J. Nutr.*, 2002 132, 3482S–3489S.

[135] Cadenas, S; Barja, G. Resveratrol, melatonin, vitamin E, and PBN protect against renal oxidative DNA damage induced by the kidney carcinogen KBrO3. *Free Radical Biol. Med.*, 1999 26, 1531-1537.

[136] Lin, JK; Tsai, S.H. Chemoprevention of cancer and cardiovascular disease by resveratrol. *Proc. Natl. Sci.* Counc. Repub. China, 1999 B23, 99-106.

[137] Tominaga, S. *Cancer incidence in Japanese in Japan, Hawaii, and western United States*. National Cancer Institute Monograph, 1969 19, 83-92.

[138] Knight, DC; Eden, JA. A review of the clinical effects of phytoestrogens. *Obstet. Gynecol.,* 1996 87, 897–904.

[139] Ziegler, R; Hoover, R; Pike, M; Hildesheim, A; Nomura, A; West, D; Wu-Williams, A; Kolonel, L; Horn-Ross, P; Rosenthal, J. Migration patterns and breast cancer risk in Asian–American women. *J. Natl. Cancer Inst.*, 1993 85, 1819–1827.

[140] Shimizu, H; Ross, RK; Bernstein, L et al. Cancers of the prostate and breast among Japanese and white immigrants in Los Angeles County. *Br. J. Cancer,* 1991 63, 963-966.

[141] Traganos, F; Ardelt, B; Halko, N; Bruno, S; Darzynkiewicz, Z. Effects of genistein on the growth and cell cycle progression of normal human lymphocytes and human leukemic MOLT-4 and HL-60 cells. *Cancer Res.*, 1992 52, 6200– 6208.

[142] Matsukawa, Y; Marui, N; Sakai, T; Satomi, Y; Yoshida, M;Matsumoto, K;Nishino, H; Aoike, A. Genistein arrests cell cycle progression at G2-M. *Cancer Res.*, 1993 53, 1328– 1331.

[143] Zhou, JR; Mukherjee, P; Gugger, ET; Tanaka, T; Blackbum, GL; Clinton, SK. Inhibition of murine bladder tumorigenesis by soy isoflavones via alterations in the cell cycle, apoptosis, and angiogenesis. *Cancer Res.*, 1998 58, 5231– 5238.

[144] Yoon, HS; Moon, SC; Kim, ND; Park, BS; Jeong, MH; Yoo, YH. Genistein induces apoptosis of RPE-J cells by opening mitochondrial PTP. *Biochem. Biophys. Res. Commun.*, 2000 276, 151–156.

[145] Baxa, DM; Luo, X; Yoshimura, FK. Genistein induces apoptosis in T lymphoma cells via mitochondrial damage. *Nutr. Cancer*, 2005 51, 93–101.

[146] Klein, CB; King, AA. Gensitein genotoxicity: Critical considerations of *in vivo* exposure dose. *Toxicol. Appl. Pharmacol.*, 2007 224, 1-11.

[147] Omenn, GS. What accounts for the association of vegetables and fruits with lower incidence of cancers and coronary heart diseases. *Ann. Epidemiol.*, 1995 5, 333– 335.

[148] Robak, J; Gryglewski, RJ. Flavonoids are scavengers of superoxide anion. *Biochem. Pharmacol.*, 1988 37, 83– 88.

[149] Torel, J; Cillard, J; Cillard, P. Antioxidant activity of flavonoids and reactivity with peroxy radical. *Phytochemistry*, 1986 25, 383–385.

[150] Husain, SR; Cillard, J; Cillard, P. Hydroxyl radical scavenging activity of flavonoids. Phytochemistry, 1987 26, 2489– 2491.

[151] Shahidi, F; Janitha, PK; Wanasundara, PD. Phenolic antioxidants. *Crit. Rev. Food Sci. Nutr.*, 1992 32, 67– 103.

[152] Barnes, S. Phytoestrogens and breast cancer. *Bail. Clin. Endocrinol. Metab.*, 1998 12, 559-666.

[153] Cassidy, A; Bingham S; Setchell, KDR. Biological effects of isoflavones in young women, importance of the chemical composition of soyabean products. *Br. J. Nutr.,* 1995 74, 587–601.

[154] Adlercreutz, H; Mousavi, Y; Clark, J; Hockerstedt, K; Hamalainen, E; Wahala, K; Makela, TH; Hase, T. Dietary phytoestrogens and cancer, in vitro and in vivo studies. *J. Steroid Biochem. Mol. Biol.,* 1992 41, 331–337.

[155] Pino, AM; Valladares, LE; Palma, MA; Mancilla, AM; Yanez, M; Albala, C. Dietary isoflavones affect sex hormone-binding globulin levels in postmenopausal women. *J. Clin. Endocrinol. Metab.*, 2000 85, 2797–2800.

[156] Berrino, F; Bellati, C; Secreto, G; Camerini, E; Pala, V; Panico, S;Allegro, G; Kaaks, R. Reducing bioavailable sex hormones through a comprehensive change in diet, the

diet and androgens (DIANA) randomized trial. *Cancer Epidemiol. Biomarkers Prev.*, 2001 10, 25–33.

[157] Hutchins, AM; Martini, MC; Olson, BA; Thomas, W; Slavin, JL. Flaxseed consumption influences endogenous hormone concentrations in postmenopausal women. *Nutr. Cancer,* 2001 39, 58–65.

[158] Xu, X; Duncan, AM; Wangen, KE; Kurzer, MS. Soy consumption alters endogenous estrogen metabolism in postmenopausal women. *Cancer Epidemiol. Biomarkers Prev.*, 2000 9, 781–786.

[159] Nagata, C; Takatsuka, N; Inaba, S; Kawakami, N; Shimizu, H. Effect of soymilk consumption on serum estrogen concentrations in premenopausal Japanese women. *J. Natl. Cancer Inst.*, 1998 90, 1830–1835.

[160] Martini, MC; Dancisak, BB; Haggans, CJ; Thomas, W; Slavin, JL. Effects of soy intake on sex hormone metabolism in premenopausal women. *Nutr. Cancer*, 1999 34, 133–139.

[161] Maskarinec, G; Williams, AE; Inouye, JS; Stanczyk, FZ; Franke, AA. A randomized isoflavone intervention among premenopausal women. *Cancer Epidemiol. Biomarkers Prev.*, 2002 11, 195–201.

[162] Nicholls, J; Lasley Bill, L; Nakajima Steven, T; Setchell, KDR; Schneeman Barbara, O. Effects of soy consumption on gonadotropin secretion and acute pituitary responses to gonadotropin-releasing hormone in women. *J. Nutr.,* 2002 132, 708–714.

[163] Xu, X; Duncan, AM; Merz, BE; Kurzer, MS. Effects of soy isoflavones on estrogen and phytoestrogen metabolism in premenopausal women. *Cancer Epidemiol. Biomarkers Prev.*, 1998 7, 1101–1108.

[164] Haggans, CJ; Hutchins, AM; Olson, BA; Thomas, W; Martini, MC; Slavin, JL. Effect of flaxseed consumption on urinary estrogen metabolites in postmenopausal women. *Nutr. Cancer,* 1999 33, 188–195.

[165] Adlercreutz, H; Bannwart, C; Wahala, K et al. Inhibition of human aromatase by mammalian lignans and isoflavonoid phytoestrogens. *J. Steroid Biochem. Mol. Biol.*, 1993 44, 147–153.

[166] Krazeisen, A; Breitling, R; Moller, G; Adamski, J. Phytoestrogens inhibit human 17beta-hydroxysteroid dehydrogenase type 5. *Mol. Cell Endocrinol.*, 2001 171, 151–162.

[167] Dai, Q; Franke, AA; Jin, F., Shu, X-O; Hebert, JR; Custer, LJ; Cheng, J; Gao, Y-T; Zheng, W. Urinary excretion of phytoestrogens and risk of breast cancer among Chinese women in Shanghai. *Cancer Epidemiol. Biomarkers Prev.*, 2002 11, 815–821.

[168] Velasquez, MT; Bhathena, SJ. Dietary phytoestrogens: a possible role in renal disease protection. *Am. J. Kidney Disease*, 2001 37, 1056-1068.

[169] Barnes, S; Peterson, G; Coward, L. Rationale for the use of genistein containing soy matrices in chemoprevention trials for breast and prostate cancer. *J. Cell Biochem.*, 1995 22, 181–187.

[170] Messina, M; Barnes, S. The role of soy products in reducing risk of cancer. *J. Natl. Cancer Inst.,* 1991 83, 541–546.

[171] Kim, H; Peterson, TG; Barnes, S. Mechanisms of action of the soy isoflavone genistein, Emerging role for its effects via transforming growth factor-β signaling pathways. *Am. J. Clin. Nutr.*, 1998 68 (suppl 6), S1418-S1425.

[172] Zhu, Y; Lin, JH-C; Liao, H-L; Verna, L; Stemerman, MB. Activation of ICAM-I promoter by lysophosphatidylcholine, Possible involvement of protein tyrosine kinases. *Biochim. Biophys. Acta.,* 1997 1345, 93-98.

[173] Zhu, Y; Lin, JH-C; Liao, H-L; Friedli, O Jr; Verna, L; Marten, NW; Straus, DS; Stemerman, MB. LDL induces transcription factor activator protein-I in human endothelial cells. *Arterioscler. Thromb. Vasc. Biol.*, 1998 18, 473-480.

[174] Barnes, PJ; Karin, M. Nuclear factor-KB, A pivotal transcription factor in chronic inflammatory diseases. *N. Engl. J. Med.,* 1997 336, 1066-1071.

[175] Lan, Q; Mercurius, KO; Davies, PF. Stimulation of transcription factor NF-KB and AP-I in endothelial cells subjected to shear stress. *Biochem. Biophys. Res. Commun.*, 1994 201, 950-956,

[176] Lamartiniere, CA; Moore, JB; Brown, NA et al. Genistein suppresses mammary cancer in rats. *Carcinogenesis*, 1995 16, 2833-2840.

[177] Adlercreutz, H; Fotsis, T; Bannwart, C et al. Determination of urinary lignans and phytoestrogen metabolites, potential antioestrogens and anticarcinogens in urine of women on various habitual diets. *J. Steroid Biochem*., 1986 25, 791–797.

[178] Pietinen, P; Stumpf, K; Mannisto, S; Kataja, V; Uusitupa, M; Adlercreutz, H. Serum enterolactone and risk of breast cancer, a case-control study in Eastern Finland. *Cancer Epidemiol. Biomarkers Prev.,* 2001 10, 339–344.

[179] Boccardo, F; Lunardi, G; Guglielmini, P; Parodi, M; Murialdo, R; Schettini, G; Rubagotti, A. Serum enterolactone levels and the risk of breast cancer in women with palpable cysts. *Eur. J. Cancer*, 2004 40, 84–89.

[180] Adlercreutz, H; Fotsis, T; Heikkinen, R et al. Excretion of the lignans enterolactone and enterodiol and of equol in omnivorous and vegetarian post-menopausal women and in women with breast cancer. *Lancet*, 1982 ii, 1295–1299.

[181] Ingram, D; Sanders, K; Kolybaba, M; Lopez, D. Case–control study of phyto-oestrogens and breast cancer. *Lancet*, 1997 350, 990–994.

[182] Thompson, LU. Experimental studies on lignans and cancer. *Bail. Clin. Endocrinol. Metab.,* 1998 12, 691–705.

[183] Yuan, JM; Wang, QS; Ross, RK et al. Diet and breast cancer in Shanghai and Tianjin, China. *Br. J. Cancer,* 1995 71, 1353–1358.

[184] Key, TJ; Sharp, GB; Appleby, PN et al. Soya foods and breast cancer risk, a prospective study in Hiroshima and Nagasaki, Japan. *Br. J. Cancer*, 1999 81, 1248–1256.

[185] Horn-Ross, PL; John, EM; Lee, M et al. Phytoestrogen consumption and breast cancer risk in a multiethnic population, the Bay Area Breast Cancer Study. *Am. J. Epidemiol.,* 2001 154, 434–441.

[186] den Tonkelaar, I; Keinan-Boker, L; Veer, PV et al. Urinary phytoestrogens and postmenopausal breast cancer risk. *Cancer Epidemiol. Biomarkers Prev.*, 2001 10, 223–228.

[187] Kilkkinen, A; Virtamo, J; Vartiainen, E; Sankila, R; Virtanen, MJ; Adlercreutz, H; Pietinen, P. Serum enterolactone concentration is not associated with breast cancer risk in a nested casecontrol study. *Int. J. Cancer,* 2004 108, 277–280.

[188] Trapman, J; Brinkmann, OA. The androgen receptor in prostate cancer. *Pathol. Res. Pract.*, 1996 192, 752–760.

[189] Lee, C; Sutkowski, DM; Sensibar, JA; Zelner, D; Kim, I; Amsel, I et al. Regulation of proliferation and production of prostate-specific antigen in androgen-sensitive prostatic cancer cells, LNCaP, by dihydrotestosterone. *Endocrinology*, 1995 136, 796–803.

[190] Russell, DW; Wilson, JD. Steroid 5 alpha-reductase, two genes/two enzymes. *Annu. Rev. Biochem.*, 1994 63, 25–61.

[191] Russell, DW; Berman, DM; Bryant, JT; Cala, KM; Davis, D.L; Landrum CP et al. The molecular genetics of steroid 5 alpha-reductases. *Recent Prog. Horm. Res.*, 1994 49, 275–284.

[192] Feldman, BJ; Feldman, D. The development of androgen independent prostate cancer. *Nat. Rev. Cancer*, 2001 1, 34– 45.

[193] Reid, P; Kantoff, P; Oh, W. Antiandrogens in prostate cancer. Invest. New Drugs, 1999 17, 271–284.

[194] Klotz, L. Hormone therapy for patients with prostate carcinoma. *Cancer*, 2000 88, 3009–3014.

[195] Welshons, WV; Thayer, KA; Judy, BM; Taylor, JA; Curran, EM; vom Saal, FS. Large effects from small exposures. I. Mechanisms for endocrine-disrupting chemicals with estrogenic activity. *Environ. Health Perspect*., 2003 111, 994–1006.

[196] Jarred, RA; Cancilla, B; Prins, GS; Thayer, KA; Cunha, GR; Risbridger, GP. Evidence that estrogens directly alter androgen-regulated prostate development. Endocrinology, 2000 141, 3471–3477.

[197] Hirayama, T. Nutrition and cancer – a large scale cohort study. *Prog. Clin. Biol. Res.*, 1986 206, 299–311.

[198] Onozawa, M; Kawamori, T; Baba, M; Fukuda, K; Toda, T; Sato H et al. Effects of a soybean isoflavone mixture on carcinogenesis in prostate and seminal vesicles of F344 rats. *Jpn. J. Cancer Res.*, 1999 90, 393–398.

[199] Landstrom, M; Zhang, JX; Hallmans, G; Aman, P; Bergh, A; Damber JE et al. Inhibitory effects of soy and rye diets on the development of dunning R3327 prostate adenocarcinoma in rats. *Prostate*, 1998 36, 151–161.

[200] Pollard, M; Wolter, W. Prevention of spontaneous prostate related cancer in lobund-wistar rats by a soy protein isolate/isoflavone diet. *Prostate*, 2000 45, 101–105.

[201] Mentor-Marcel, R; Lamartiniere, CA; Eltoum, IE; Greenberg, NM; Elgavish, A. Genistein in the diet reduces the incidence of poorly differentiated prostatic adenocarcinoma in transgenic mice (TRAMP). *Cancer Res*., 2001 61, 6777–6782.

[202] Fritz, WA; Wang, J; Eltoum, IE; Lamartiniere, CA. Dietary genistein down-regulates androgen and estrogen receptor expression in the rat prostate. *Mol. Cell. Endocrinol.*, 2002 186, 89–99.

[203] Jang, M; Pezzuto, JM. Cancer chemopreventive activity of resveratrol. *Drugs Exp. Clin. Res*., 1999 25, 65–77.

[204] Landstrom, M; Zhang, JX; Hallmans, G; Aman, P; Bergh, A; Damber, JE et al. Inhibitory effects of soy and rye diets on the development of Dunning R3327 prostate adenocarcinoma in rats. *Prostate*, 1998 1–36, 151–161.

[205] Bylund, A; Zhang, JX; Bergh, A; Damber, JE; Widmark, A; Johansson, A et al. Rye bran and soy protein delay growth and increase apoptosis of human LNCaP prostate adenocarcinoma in nude mice. *Prostate*, 2000 42, 304–314.

[206] Sarkar, FH; Li, Y. Mechanisms of cancer chemoprevention by soy isoflavone genistein. *Cancer Metastasis Rev*., 2002 21, 265–280.

[207] Castle, EP; Thrasher, JB. The role of soy phytoestrogens in prostate cancer. *Urol. Clin. North Am.*, 2002 297, 1–81.

[208] Adlercreutz, H; Mazur, W; Bartels, P; Elomaa, V; Watanabe, S; Wahala, K et al. Phytoestrogens and prostate disease. *J. Nutr.*, 2000 130, 658S–659S.

[209] Hussain, M; Banerjee, M; Sarkar, FH; Djuric, Z; Pollak, MN; Doerge D et al. Soy isoflavones in the treatment of prostate cancer. *Nutr. Cancer,* 2003 47, 111–117.

[210] Hess-Wilson, JK; Knudsen, KE. Endocrine disrupting compounds and prostate cancer. *Cancer Let.*, 2006 241, 1-12.

[211] Nagata, C; Inaba, S; Kawakami, N. Inverse association of soy product intake with serum androgen and estrogen concentrations in Japanese men. *Nutr. Cancer*, 2000 36, 14–18.

[212] Kuiper, GG; Carlsson, B; Grandien, K. Cloning of a novel estrogen receptor expressed in a rat prostate and ovary. *Proc. Natl. Acad. Sci.,* 1996 93, 5925–5930.

[213] Barnes, S. Role of phytochemicals in prevention and treatment of prostate cancer. *Epidemiol. Rev.*, 2001 23, 102–105.

[214] Demark-Wahnefried, W; Price, DT; Polascik, TJ; Robertson, CN; Anderson, EE; Paulson, DF et al. Pilot study of dietary fat restriction and flaxseed supplementation in men with prostate cancer before surgery, exploring the effects on hormonal levels, prostate-specific antigen, and histopathologic features. *Urology*, 2001 58, 47–52.

[215] Weber, KS; Setchell, KD; Stocco, DM; Lephart, ED. Dietary soy-phytoestrogens decrease testosterone levels and prostate weight without altering LH, prostate 5 alpha-reductase or testicular steroidogenic acute regulatory peptide levels in adult male sprague-dawley rats. *J. Endocrinol.,* 2001 170, 591–599.

[216] Kato, K; Takahashi, S; Cui, L; Toda, T; Suzuki, S; Futakuchi M et al. Suppressive effects of dietary genistin and daidzin on rat prostate carcinogenesis. *Jpn. J. Cancer Res.*, 2000 91, 786–791.

[217] Hempstock, J; Kavanagh, JP; George, NJ. Growth inhibition of prostate cell lines in vitro by phyto-oestrogens. *Br. J. Urol.*, 1998 82, 560–563.

[218] Santibanez, JF; Navarro, A; Martinez, J. Genistein inhibits proliferation and in vitro invasive potential of human prostatic cancer cell lines. *Anticancer Res.*, 1997 17, 1199–1204.

[219] Lee, MM; Wang, RT; Hsing, AW et al. Case–control study of diet and prostate cancer in China. *Cancer Causes Contro*l, 1998 9, 545–552.

[220] Oishi, K; Okada, K; Yoshida, O et al. A case–control study of prostatic cancer with reference to dietary habits. *Prostate*, 1988 12, 179–190.

[221] Hirayama, T. *Epidemiology of prostate cancer with special reference to the role of diet.* National Cancer Institute Monograph, 1979 53, 149–155.

[222] Nagata, C; Shimizu, H; Takami, R; Hayashi, M; Takeda, N; Yasuda, K. Soy product intake and hot flashes in Japanese women. *Am. J. Epidemiol.*, 2001 153, 790–793.

[223] Han, KK; Soares, JM Jr; Haidar, MA; de Lima, GR; Baracat, EC. Benefits of soy isoflavone therapeutic regimen on menopausal symptoms. *Obstet. Gynecol.,* 2002 99, 389–394.

[224] Krebs, EE; Ensrud, KE; MacDonald, R; Wilt, TJ. Phytoestrogens for treatment of menopausal symptoms, a systematic review. *Obstet. Gynecol.*, 2004 104, 824–836.

[225] Baber, RJ; Templeman, C; Morton, T; Kelly, GE; West, L. Randomized placebo-controlled trial of an isoflavone supplement and menopausal symptoms in women. *Climacteric*, 1999 2, 85–92.

[226] Messina, M; Hughes, C. Efficacy of soyfoods and soybean isoflavone supplements for alleviating menopausal symptoms is positively related to initial hot flush frequency. *J. Med. Food,* 2003 6, 1–11.

[227] Barnes S. Phyto-oestrogens and osteoporosis, what is a safe dose? *Br. J. Nutr.* 2003 89, 101–108.

[228] Tempfer, CB; Bentz, E-K; Leodolter, S; Tscherne, G; Reuss, F; Cross, HS; Huber, JC. Phytoestrogens in clinical practice: a review of the literature. *Fertil. Steril.,* 2007 87, 1243-1249.

[229] Albertazzi, P; Pansini, F; Bonaccorsi, G; Zanotti, L; Forini, E; De Aloysio, D. The effects of soy supplementation on hot flushes. *Obstet. Gynecol.*, 1998 91, 6 –11.

[230] Dalais, FS; Rice, GE; Wahlqvist, ML et al. Effects of dietary phytoestrogens in post-menopausal women. *Climacteric*, 1998 1, 124–129.

[231] van Patten, CL; Olivotto, IA; Chambers, GK et al. Effect of soy phytoestrogens on hot flushes in postmenopausal women with breast cancer, a randomized, controlled clinical trial. *J. Clin. Oncol.*, 2002 20, 1449–1455.

[232] Brzezinski, A; Adlercreutz, H; Shaoul, R et al. Short-term Effects of phytoestrogen-rich diet on postmenopausal women. *J. bNorth Am. Menopause Soc.*, 1997 4, 89–94.

[233] Scambia, G; Mango, D; Signorile, PG et al. Clinical effects of a standardazided soy extract in postmenopausal women, a pilot study. *Menopause*, 2000 2, 150–151.

[234] Baber, RJ; Templeman, C; Morton, T; Kelly, GE; West L. Randomised placebo-controlled trial of an isoflavone supplement and menopausal symptom in women. *Climacteric*, 1999 2, 85–92.

[235] Yamaguchi, M. Isoflavone and bone metabolism, its cellular mechanism and preventive role in bone loss. *J. Health Sci.*, 2002 48, 209–222.

[236] Hays, J; Ockene, JK; Brunner, RL; Kotchen, JM; Manson, JE; Patterson, RE; Aragaki, AK; Shumaker, SA; Brzyski, RG; LaCroix, AZ; Granek, IA; Valanis, BG., Women's Health Initiative Investigators. Effects of estrogen plus progestin on health related quality of life. *N. Engl. J. Med.*, 2003 348, 1835–1837.

[237] Lauderdale, DS; Jacobsen, SJ; Furner, SE et al. Hip fracture incidence among elderly Asian–American populations. *Am. J. Epidemiol.*, 1997 146, 502–509.

[238] Alekel, DL; Germain, AS; Peterson, CT; Hanson, KB; Stewart, JW; Toda, T. Isoflavone-rich soy protein isolate attenuates bone loss in the lumbar spine of perimenopausal women. *Am. J. Clin. Nutr.*, 2000 72, 844–852.

[239] Mei, J; Yeung, SS; Kung, A.W. High dietary phytoestrogen intake is associated with higher bone mineral density in postmenopausal but not premenopausal women. *J. Clin. Endocrinol. Metab.*, 2001 86, 5217–5221.

[240] Morabito, N; Crisafulli, A; Vergara, C; Gaudio, A; Lasco, A; Frisina, N; D'Anna, R; Corrado, F; Pizzoleo, MA; Cincotta, M; Altavilla, D; Ientile, R; Squadrito, F. Effects of genistein and hormone-replacement therapy on bone loss in early postmenopausal women, a randomized double-blind placebo-controlled study. *J. Bone Mineral Res.*, 2002 17, 1904–1912.

[241] Potter, S; Baum, J; Teng, H; Stillman, R; Shay, N; Erdman Jr, J. Soy protein and isoflavones, their effects on blood lipids and bone density in postmenopausal women. *Am. J. Clin. Nutr.*, 1998 68, 1375-1379.

[242] Cross, HS; Kallay, E; Lechner, D; Gerdenitsch, W; Adlercreutz, H; Armbrecht, HJ. Phytoestrogens and vitamin D metabolism, a new concept for the prevention and therapy of colorectal, prostate and mammary carcinomas. *J. Nutr.*, 2004 134, 1207–1212.

[243] Gennari, C; Agnusdei, D; Crepaldi, G et al. Effect of ipriflavone—a synthetic derivative of natural isoflavones—on bone mass loss in the early years after menopause. *Menopause*, 1998 5, 9–15.

[244] Yoshida, K; Tsukamoto, T; Torii, H et al. Metabolism of ipriflavone (TC-80) in rats. *Radioisotopes*, 1985 34, 612–617.

[245] Gambacciani, M; Spinetti, A; Piaggesi, L et al. Ipriflavone prevents the bone mass reduction in premenopausal women treated with gonadotropin hormone-releasing hormone agonists. *Bone Mineral,* 1994 26, 19–26.

[246] Tsutsumi, N. Effect of coumestrol on bone metabolism in organ culture. *Biol. Pharm. Bull.*, 1995 18, 1012-1015.

[247] Akiyama, T; Ishida, J; Nakagawa, S et al. Genistein, a specific inhibitor of tyrosine-specific protein kinases. *J. Biol. Chem.*, 1987 262, 5592-5595.

[248] Okura, A; Arakawa, H; Oka, H et al. Effect of genistein on topoisomerase activity and the growth of val 12 H-ras-transformed NIH3T3 cells. *Biochem. Biophy. Res. Comm.,* 1988 157, 183-189.

[249] Watson, CS; Pappas, TC; Gametchu, B. The other estrogen receptor in the plasma membrane, implications for the actions of environmental estrogens. *Environ. Health Persp.*, 1995 103 (suppl. 7), 41-50.

[250] Williams, JE; Jordan, SE; Barnes, S; Blair, HC. Tyrosine kinase inhibitor effects on osteoclastic acid transport. *Am. J.Clin. Nutr.*, 1998 68, 13 695 13 745.

[251] Kreijkamp-Kaspers, S; Kok, L; Grobbee, DE; de Haan, EH; Aleman, A; Lampe, JW; van der Schouw, YT. Effect of soy protein containing isoflavones on cognitive function, bone mineral density, and plasma lipids in postmenopausal women, a randomized controlled trial. *JAMA*, 2004 292, 65–74.

[252] Beaglehole, R. International treands in coronary heart disease mortality, morbidity and risk factors. *Epidemiol. Rev.*, 1990 12, 1–15.

[253] Rubanyi, GM; Johns, A; Kauser, K. Effect of estrogen on endothelial function and angiogenesis. *Vasc. Pharmacol.*, 2002 38, 89–98.

[254] Washburn, S; Burke, GL; Morgan, T; Anthony, M. Effect of soy protein supplementation on serum lipoproteins, blood pressure, and menopausal symptoms in perimenopausal women. *Menopause*, 1999 6, 7–13.

[255] Merz-Demlow, BE; Duncan, AM; Wangen, KE; Xu, X; Carr, TP; Phipps, WR; Kurzer, MS. Soy isoflavones improve plasma lipids in normocholesterolemic, premenopausal women. *Am. J. Clin. Nutr.*, 2000 71, 1462–1469.

[256] Lucas, EA; Wild, RD; Hammond, LJ; Khalil, DA; Juma, S; Daggy, BP; Stoecker, BJ; Arjmandi, BH. Flaxseed improves lipid profile without altering biomarkers of bone metabolism in postmenopausal women. J. *Clin. Endocrinol. Metab.,* 2002 87, 1527–1532.

[257] Cruz, ML; Wong, WW; Mimouni, F; Hachey, DL; Setchell, KD; Klein, PD; Tsang, RC. Effects of infant nutrition on cholesterol synthesis rates. *Pediatr. Res.*, 1994 35, 135–140.

[258] Anderson, JW; Johnstone, BM; Cook-Newell, ME. Metaanalysis of the effects of soy protein intake on serum lipids. *New Engl. J. Med.*, 1995 333, 276–282.

[259] Sirtori, CR; Galli, G; Lovati, MR; Cxarrara, P; Bosisio, E; Kienle, MG. Effects of dietary proteins on regulation of liver lipoprotein receptors in rats. *J. Clin. Nutr.*, 1998 114, 1493–1500.

[260] Baum, JA; Teng, H; Erdman, JW et al. Long-term intake of soy protein improves blood lipids profiles and increases mononuclear cell LDL receptor mRNA in hypocholesterolemic postmenopausal women. *Am. J. Clin. Nutr.*, 1998 68, 545–551.

[261] Jenkins, DJ; Kendall, CW; Vidgen, E et al. Health aspects of partially defatted flaxseed, including effects on serum lipids, oxidative measures, and in vivo androgen and progestin activity, a controlled crossover trial. *Am. J. Clin. Nutr.*, 1999 69, 395–402.

[262] Witztum, JL; Steinberg, D. Role of oxidized low density lipoprotein in atherogenesis. *J. Clin. Inv.*, 1991 88, 1785-1792.

[263] Simoncini, T; Fornari, L; Mannella, P; Caruso, A; Garibaldi, S; Baldacci, C; Genazzani AR. Activation of nitric oxide synthesis in human endothelial cells by red clover extracts. *Menopause*, 2005 12, 69–77.

[264] Gottstein, N; Ewins, BA; Eccleston, C; Hubbard, GP; Kavanagh, IC; Minihane, AM et al. Effect of genistein and daidzein on platelet aggregation and monocyte and endothelial function. *Br. J. Nutr.*, 2003 89, 607–616.

[265] Howes, JB; Sullivan, D; Lai, N; Nestel, P; Pomeroy, S; West, L; Eden, JA; Howes, LG. The effects of dietary supplementation with isoflavones from red clover on the lipoprotein profiles of post menopausal women with mild to moderate hypercholesterolaemia. *Atherosclerosis*, 2000 152, 143–147.

[266] Dewell, A; Hollenbeck Clarie, B; Bruce, B. The effects of soy-derived phytoestrogens on serum lipids and lipoproteins in moderately hypercholesterolemic postmenopausal women. *J. Clin. Endocrinol. Metab.*, 2002 87, 118–121.

[267] De Kleijn, MJJ; Van der Schouw, YT; Wilson, PWF; Grobbee, DE; Jacques, PF. Dietary intake of phytoestrogens is associated with a favorable metabolic cardiovascular risk profile in postmenopausal US. women, the Framingham Study. *J. Nutr.*, 2002 132, 276–282.

[268] Van der Schouw, YT; Pijpe, A; Lebrun, CEI; Bots, ML; Peeters, PHM; Van Straveren, WA; Lamberts, SWJ; Grobbee, DE. Higher usual dietary intake of phytoestrogens is associated with lower aortic stiffness in postmenopausal women. Arterioscler. *Thromb. Vasc. Biol.*, 2002 22, 1316–1322.

[269] Fotsis, T; Pepper, MS; Montesano, R; Aktas, E; Breit, S; Schweigerer, L; Rasku, S; Wähälä, K; Adlercreutz, H. Phtoestrogens and inhibition of angiogenesis. *Bail. Clin. Endocrinol. Metab.*, 1998 12, 649-666.

[270] Denekamp, J. Vascular attack as a therapeutic strategy for cancer. Cancer Metab. Rev., 1990 9, 267-282.

[271] Folkman, J. Tumor angiogenesis. *Adv. Cancer Res.*, 1985 43, 175-203.

[272] Pepper, MS; Montesano, R. Proteolytic balance and capillary morphogenesis. *Cell Differentiation and Development,* 1990 32, 319-328.

[273] Folkman, J; Klagsbrun, M. Angiogenic factors. Science, 1987 235, 442-447.
[274] Montesano, R; Vassalli, J-D; Baird, A et al. Basic fibroblast growth factor induces angiogenesis in vitro. *Proc. Nat. Acad. Sci.* USA, 1986 83, 7297-7301.
[275] Fotsis, T; Pepper, M; Adlercreutz, H et al. Genistein a dietary-derived inhibitor of *in vitro* angiogenesis. *Proc. Nat. Acad. Sci.* USA, 1993 90, 2690-2694.
[276] Byers, M; Kuiper, GGJM; Gustafsson, JA; Park-Sarge, OK. Estrogen receptor-mRNA expression in rat ovary, down-regulation by gonadotropins. *Mol. Endocrinol.*, 1997 11, 172-182.
[277] Wang, W; Tanaka, Y; Hau, Z; Higuchi, CM. Proliferative response of mammary glandular tissue to formononetin. *Nutr. Cancer,* 1995 23, 131-140.
[278] Santell, RC; Chang, YC; Nair, MG; Helferich, WG. Dietary genistein exerts estrogenic effects upon the uterus, mammary gland, and the hypothalamic/pituitary axis in rats. *J. Nutr.*, 1997 127, 263-269.
[279] Cassidy, A. Phytoestrogens, a possible alternative to HRT? J. *Br. Menopause Soc.,* 1998 suppl. SI, 10-11.
[280] Cassidy, A; Bingham, S; Setchell, KDR. Biological effects of a diet of soy protein rich in isoflavones on the menstrual cycle of premenopausal women. *Am. J. Clin. Nutr.,* 1994 60, 333-340.
[281] Coldham, NG; Dave, M; Sivapathasundaram, S et al. Evaluation of a recombinant yeast cell estrogen screening assay. *Environ. Health Perspectives*, 1997 105, 734-742.
[282] Dodge, JA; Glasebrook, AL; Magee, DE et al. Environmental estrogens, effects on cholesterol lowering and bone in the ovariectomized rat. *J. Steroid Biochem. Mol. Biol.*, 1996 59, 155-161.
[283] Nwannenna, AI; Madej, A; Lundh, TJO; Fredriksson, G. Effects of oestrogenic silage on some clinical and endocrinological parameters in ovariectomized heifers. *Acta Veterinaria Scandinavica*, 1994 35, 173.
[284] Tansey, G; Hughes, CLJ; Cline, JM et al. Effects of dietary soybean estrogens on the reproductive tract in female rats. *Proc. Soc. Exp. Biol. Med.,* 1998 217, 340-344.
[285] Clarkson, TB; Anthony, MS; Williams, JK et al. The potential of soybean phytoestrogens for postmenopausal hormone replacement therapy. *Proc. Soc. Exp. Biol. Med.*, 1998 217, 365-368.
[286] Cline, JM; Paschold, JC; Adams, MS et al. Effects of hormonal therapies and dietary soy phytoestrogens on vaginal cytology in surgically postmenopausal macaques. *Fertil. Steril.*, 1996 65, 1031-1035.
[287] Stammel, W; Thomas, H; Staib, W; Kuhn-Velten, WN. Tetrahydroisoquinoline alkaloids mimic direct but not receptor-mediated inhibitory effects of estrogens and phytoestrogens on testicular endocrine function. Possible significance for Leydig insufficiency in alcohol addiction. *Life Sciences*, 1991 49, 1319-1329.
[288] Anthony, MS; Clarkson, TB; Bullock, BC; Wagner, JD. Soy protein versus soy phyotestrogens in the prevention of diet-induced coronary artery atherosclerosis of male cynomologus monkeys. Arterioscler. *Thromb. Vasc. Biol.,* 1997 17, 2524-2531.
[289] Anthony, MS; Clarkson, TB; Hughes, CL Jr et al. Soybean isoflavones improve cardiovascular risk factors without affecting the reproductive system of peripubertal rhesus monkeys. *J. Nutr.,* 1996 126, 43-50.
[290] Whitten, PL; Naftolin, F. *Xenoestrogens and neuroendocrine development.* In: Needleman, HL; Bellinger, D, editor. Prenatal Exposure to Environmental Toxicants,

Developmental Consequences. Baltimore: Johns Hopkins University Press, 1994, 269-293.
[291] Toppari, J; Larsen, JC; Christiansen, P et al. Male reproductive health and environmental xenoestrogens. *Environ. Health Perspectives*, 1996 104, 741-803.
[292] Gallo, D; Cantelmo, F; Distefano, M; Ferlini, C; Zannoni, GF; Riva, A; Morazzoni, P; Bombardelli, E; Mancuso, S; Scambia, G. Reproductive effects of dietary soy in female Wistar rats. *Food Chem. Toxicol.*, 1999 37, 493-502.
[293] Murphy, PA; Song, TT; Buseman, G; Barua, K. Isoflavones in soy-based infant formulas. *J.Agric. Food Chem.*, 1997 45, 4635-4638.
[294] Setchell, KDR; Zimmer-Nechemias, L; Cai, J; Heubi, JE. Exposure of infants to phytooestrogens from soy-based infant formula. *Lancet*, 1997 350, 23-27.
[295] Barnes, S. Effect of genistein on in vitro and in vivo models of cancer. *J. Nutr.*; 1995 125, S777.
[296] Levy, JR; Faber, KA; Ayyash, L; Hughes, CL Jr. The effect of prenatal exposure to the phytoestrogen genistein on sexual differentiation in rats. *Proc. Soc. Exp. Biol. Med.*, 1995 208, 60-66.
[297] McClain, RM; Wolz, E; Davidovich, A; Edwards, J; Bausch, J. Reproductive safety studies with genistein in rats. *Food Chem. Toxicol.*, 2007 45, 1319-1332.
[298] Medlock, KL; Branham, WS; Sheehan, DM. Effects of coumestrol and equol on the developing reproductive tract of the rat. *Proc. Soc. Exp. Biol. Med.,* 1995 208, 67-71.
[299] Whitten, PL; Lewis, C; Naftolin, F. A phytoestrogen diet induces the premature anovulatory syndrome in lactationally exposed female rats. *Biol. Reprod.*, 1993 49, 1117-1121.
[300] Burroughs, CD. Long-term reproductive tract alterations in female mice treated neonatally with coumestrol. *Proc. Soc. Exp. Biol. Med.*, 1995 208, 78-81.
[301] Russo, J; Russo, IH. Biological and molecular bases of mammary carcinogenesis. *Lab. Investigations,* 1987 57, 112-137.
[302] Lopez, J; Ogren, L; Verjan, R; Talamantes, F. Effects of perinatal exposure to a synthetic estrogen and progestin on mammary tumorigenesis in mice. *Teratology*, 1988 38, 129-134.
[303] Murrill, WB; Brown, NM; Zhang, J et al. Prepubertal genistein exposure suppresses mammary cancer and enhances gland differentiation in rats. *Carcinogenesis*, 1997 17, 1451.
[304] Lamartiniere, CA; Murrill, WB; Manzolillo, PA et al. Genistein alters the ontogeny of mammary gland development and protects against chemically-induced mammary cancer in rats. *Proc. Soc. Exp. Biol. Med.*, 1998 217, 358-364.
[305] Pan, Y; Anthony, M; Clarkson, TB. Effect of estradiol and soy phytoestrogens on choline acetyltransferase and nerve growth factor mRNAs in the frontal cortex and hippocampus of female rats. *Proc. Soc. Exp. Biol. Med.*, 1999 221, 118–125.
[306] Pan, Y; Anthony, M; Clarkson, TB. Evidence for up-regulation of brain-derived neurotrophic factor mRNA by soy phytoestrogens in the frontal cortex of retired breeder female rats. *Neurosci. Lett.*, 1999 261, 17–20.
[307] Kim, H; Xia, H; Li, L; Gewin, J. Attenuation of neurodegeneration-relevant modifications of brain proteins by dietary soy. Biofactors, 2000 12, 243–250.
[308] Gamache, PH; Maher, TJ; Setchell, KDR; Wu, T-H; Acworth, IN. The transfer of non-steroidal dietary estrogens into brain. *Soc. Neurosci. Abstr.*, 1996 22, 1967.

[309] Arnold, AP; Gorski, RA. Gonadal steroid interaction of structural sex differences in the central nervous system. Annu. Rev. Neurosci., 1984 7, 413–422.

[310] Davis, SR; Shryne, JE; Gorski, RA. A revised critical period for the sexual differentiation of the sexually dimorphic nucleus of the preoptic area in the rat. *Neuroendocrinology*, 1995 62, 579–585.

[311] Simerly, RB. Organization and regulation of sexually dimorphic neuroendocrine pathways. *Behav. Brain Res.*, 1998 92, 195–203.

[312] De Jonge, FH; Louwerse, AL; Ooms, P; Evers, MP; Endert, E; van de Poll, NE. Lesions of the SDN-POA inhibit sexual behavior of Wistar rats. *Brain Res. Bull.*, 1987 23, 483–492.

[313] Faber, KA; Hughes Jr., CL. The effect of neonatal exposure to diethylstilbestrol, genistein, and zearalenone on pituitary responsiveness and sexually dimorphic nucleus volume in the castrated adult rat. *Biol. Reprod.*, 1991 45, 649–653.

[314] Faber, KA; Hughes Jr., CL. Dose–response characteristics of neonatal exposure to genistein on pituitary responsiveness to gonadotropin releasing hormone and volume of the sexually dimorphic nucleus of the preoptic area (SDN-POA) in postpubertal castrated female rats. *Reprod. Toxicol.,* 1993 7, 35–39.

[315] Lephart, ED; Rhees, RW; Setchell, KDR; Bu, LH; Lund, TD. Estrogens and phytoestrogens, brain plasticity of sexually dimorphic brain volumes. *J. Steroid Biochem. Mol. Biol.*, 2003 85, 299–309.

[316] Celotti, F; Melcangi, C; Martini, L. The 5a-reductase in the brain, Molecular aspects and relation to brain function. *Front. Neuroendocrinol.*, 1992 13, 163–215.

[317] Celotti, F; Negri-Cesi, P; Poletti, A. Steroid metabolism in the mammalian brain, 5 alpha-reduction and aromatization. *Brain Res. Bull.*, 1997 44, 365–375.

[318] Lephart, ED. A review of brain 5a-reductase: Cellular, enzymatic and molecular perspectives and implications for biological function. *Mol. Cell Neurosci.*, 1993 4, 473–484.

[319] Lephart, ED. A review of brain aromatase cytochrome P450. *Brain Res. Rev.*, 1996 22, 1–26.

[320] MacLusky, NJ; Naftolin, F. Sexual differentiation of the central nervous system. *Science*, 1981 211, 1294–1302.

[321] Roselli, CE; Horton, LE; Resko, JA. Distribution and regulation of aromatase activity in the rat hypothalamus and limbic system. *Endocrinology*, 1985 117, 2471–2476.

[322] Lephart, ED. A review of brain 5a-reductase, cellular, enzymatic, and molecular perspectives and implications for biological function. *Mol. Cell. Neurosci.*, 1993 4, 473–484.

[323] Naftolin, F. Brain aromatization of androgens. J. Reprod. Med., 1994 39, 257– 261.

[324] Garcia-Segura, LM; Azcoitia, I; DonCarlos, LL. Neuroprotection by estradiol. *Prog. Neurobiol.*, 2001 63, 29–60.

[325] Simpkins, JW; Green, PS; Gridley, KE; Singh, M; de Fiebre, NC; Rajakumar, G. Role of estrogen replacement therapy in memory enhancement and the prevention of neuronal loss associated with Alzheimer's disease. *Am. J. Med.*, 1997 103, 19S–25S.

[326] Fink, G; Sumner, BE; McQueen, JK; Wilson, H; Rosie, R. Sex steroid control of mood, mental state and memory. *Clin. Exp. Pharmacol. Physiol.*, 1998 25, 764–775.

[327] Sherwin, BB. Estrogen and cognitive function in women. *Proc. Soc. Exp. Biol. Med.*, 1998 217, 17–22.

[328] Martini, L; Melcangi, RC; Maggi, R. Androgen and progesterone metabolism in the central and peripheral nervous system. J. Steroid Biochem. *Mol. Biol.,* 1993 47, 195–205.
[329] Negri-Cesi, P; Poletti, A; Celotti, F. Metabolism of steroids in the brain: a new insight into the role of 5 alpha-reductase and aromatase in brain differentiation and functions. J. Steroid Biochem. *Mol. Biol.,* 1996 58, 455–466.
[330] MacLusky, NJ; Clark, CR; Shanabrough, M; Naftolin, F. Metabolism and binding of androgens in the spinal cord of the rat. *Brain Res.*, 1987 422, 83–91.
[331] Lephart, ED; Ladle, DR; Jacobson, NA; Rhees, RW. Inhibition of brain 5 alpha-reductase in pregnant rats, effects on enzymatic and behavioral activity. *Brain Res.,* 1996 739, 356–360.
[332] Weber, KS; Jacobson, NA; Setchell, KDR; Lephart, ED. Brain aromatase and 5α-reductase, regulatory behaviors and testosterone levels in adult rats on phytoestrogen diets. *Proc. Soc. Exp. Biol. Med.*, 1999 22, 131–135.
[333] Iacopino, AM; Quintero, EM; Miller, EK. Calbindin-D a 28-K potential neuroprotective protein. *Neurodegeneration*, 1994 3, 1–20.
[334] Lephart, ED; Thompson, JM; Setchell, KDR; Adlercreutz, H; Weber, KS. Phytoestrogens decrease brain calbindin-binding proteins but do not alter hypothalamic androgen metabolizing enzymes in adult male rats. *Brain Res.*, 2000 859, 123–131.
[335] Taylor, H; Quintero, EM; Iacopino, AM; Lephart, ED. Phytoestrogens alter hypothalamic calbindin-D28K levels during prenatal development. *Dev. Brain Res.*, 1999 114, 277–281.
[336] D'Amico, G; Gentile, MG; Manna, G; Fellin, G; Ciceri, R; Cofano, F; Petrini, C; Lavarda, F; Perolini, S; Porrini, M. Effect of vegetarian soy diet on hyperlipidaemia in nephritic syndrome. *Lancet*, 1992 339, 1131-1134.
[337] Gentile, MG; Fellin, G; Cofano, E; Delle Fave, A; Manna, G; Ciceri, R; Petrini, C; Lavarda, E; Pozzi, F; D'Amico, G. Treatment of proteinuric patients with a vegetarian soy diet and fish. *Clin. Nephrol.*, 1993 40, 315-320.
[338] Clark, WE; Parbtani, A; Huff, MW; Spanner, E; de Salis, H; Chin-Yee, I; Philbrick, DJ; Holub, BJ. Flaxseed, A potential treatment for lupus nephritis. *Kidney Int.*, 1995 48, 475-480.
[339] Pelkmans, L; Helenius, A. Insider information, what viruses tell us about endocytosis. *Curr. Opin. Cell. Biol.,* 2003 15, 414–422.
[340] Smith, AE; Helenius, A. How viruses enter cells. Science, 2004 304, 237– 242.
[341] Pelkmans, L; Helenius, A. Endocytosis via caveolae. *Traffic*, 2002 3, 311–320.
[342] Pelkmans, L; Puntener, D; Helenius, A. Local actin polymerization and dynamin recruitment in SV40-induced internalization of caveolae. *Science*, 2002 296, 535–539.
[343] Dangoria, NS; Breau, WC; Anderson, HA; Cishek, DM; Norkin, LC. Extracellular simian virus 40 induces an ERK/MAP kinase-independent signaling pathway that activates primary response genes and promotes virus entry. *J. Gen. Virol.*, 1996 77, 2173–2182.
[344] Hirasawa, K; Kim, A; Han, HS; Han, J; Jun, HS; Yoon, JW. Effect of p38 mitogen-activated protein kinase on the replication of encephalomyocarditis virus. *J. Virol.*, 2003 77, 5649–5656.
[345] Talavera, D; Castillo, AM; Dominguez, MC; Gutierrez, AE; Meza, I. IL8 release, tight junction and cytoskeleton dynamic reorganization conducive to permeability increase

are induced in dengue virus infection of microvascular endothelial monolayers. *J. Gen. Virol.*, 2004 85, 1801–1813.

[346] Cirone, M; Zompetta, C; Tarasi, D; Frati, L; Faggioni, A. Infection of human T lymphoid cells by human herpesvirus 6 blocked by two unrelated protein tyrosine kinase inhibitors, biochanin A and herbimycin. AIDS Res. *Hum. Retroviruses*, 1996 12, 1629–1634.

[347] Greiner, LL; Stahly, TS; Stabel, TJ. The effect of dietary soy genistein on pig growth and viral replication during a viral challenge. *J. Anim. Sci.*, 2001 79, 1272–1279.

[348] Padilla-Banks, E; Jefferson, WN; Newbold, RR. Neonatal exposure to the phytoestrogen genistein alters mammary gland growth and developmental programming of hormone receptor levels. *Endocrinology*, 2006 147, 4871–4882.

[349] McClain, RM; Wolz, E; Davidovich, A; Bausch, J. Genetic toxicity studies with genistein. *Food Chem. Toxicol.*, 2006 44, 42–55.

[350] Snyder, RD; Gillies, PJ. Reduction of genistein clastogenicity in Chinese hamster V79 cells by daidzein and other flavonoids. *Food Chem. Toxicol.*, 2003 41, 1291–1298.

[351] Merritt, RJ; Jenks, BH. Safety of soy-based infant formulas containing isoflavones, the clinical evidence. *J. Nutr.*, 2004 134, 1220S–1224S.

[352] Tuohy, P. Soy infant formula and phytoestrogens (Review article). *J. Paediatr. Child Health,* 2003 39, 401–405.

[353] Mazur, WM; Adlercreutz, H. Naturally occurring oestrogens in food. *J. Pure Appl. Chem.*, 1998 70, 1759-1776.

Ligands, Polymers and Amino Acids
Editor: Tsisana Shartava

ISBN: 978-1-61122-793-2

THE STRUCTURE OF α_1-PROTEINASE INHIBITOR POLYMER: FACTS AND HYPOTHESES

Ewa Marszal

Division of Hematology, Office of Blood Research and Review, Center for Biologics Evaluation and Research, Food and Drug Administration, Bethesda, MD, USA

ABSTRACT

The metastable structure of a serpin α_1-proteinase inhibitor (α_1-PI) makes it prone to conformational changes as a result of certain mutations and under mild denaturing conditions. Polymers of several variants of α_1-PI, e.g., the most clinically relevant Z variant, accumulate in the liver and are believed to be the underlying cause of the liver disease associated with α_1-PI deficiency. Such polymers have also been found in the circulation and in the lung. The structure of these polymers, which is important for structure-based drug design, remains unknown. The historically first and generally accepted model of the α_1-PI polymers, the loop-A sheet model, which assumes insertion of the reactive center loop (RCL) of one α_1-PI molecule between the central strands of the A β-sheet of another molecule, has never been proven. In addition, this model is inconsistent with resonance energy transfer data obtained for heteropolymers formed from the Z and S variants. It is also difficult to envision how this model or even the more compact loop-A sheet model, deduced from fluorescence data obtained for the Z and S variant heteropolymers, can be compatible with electron microscopy (EM) images, which show apparently flexible polymer chains. Two other polymer models were deduced from the interactions seen between molecules in the crystals of two other serpins, antithrombin (AT) and plasminogen activator inhibitor-1 (PAI-1), which demonstrated the ability of the reactive center loop (RCL) to assume the β-strand conformation and to extend the C or A β-sheet in a neighboring molecule by formation of an additional edge strand. This became the ground for the loop-C sheet and strand s7A models of the polymer, although the interactions seen in crystals appear not to be stable in solution and, thus, appear not to reflect interactions existing in α_1-PI polymers, which actually are stable. The focus of this paper is to review the observations used to support the proposed models and to revisit the wealth of literature data in search for unexplained observations and possible new interpretations. I also put forward the hypothesis that a fusion of β-sheets, rather than insertion of the reactive center loop into a β-sheet, may underlie the polymerization process of α_1-PI. This hypothesis is an expansion of the recent head-to-head model proposed based on the polymerization of the α_1-PI disulfide-linked dimer.

Introduction

α_1-PI, which modulates the activity of neutrophil elastase in the lung, is a member of a large family of serine protease inhibitors (serpins) [Irving et al., 2000] that share a highly conserved structure (Figure 1). The native structure of a serpin is trapped in a metastable state, which is crucial for its function but also makes it sensitive to destabilizing mutations that may cause misfolding and polymerization and thus, may lead to a conformational disease [Stein and Carrell, 1995, Cho et al., 2005]. Retention of the Z-variant (E342K) in hepatocytes, the major site of α_1-PI synthesis, due to polymerization, leads to reduced levels of α_1-PI in the circulation, lung emphysema, and the increased risk of liver disease, which may have a range of clinical presentations from neonatal hepatitis to liver cirrhosis and hepatocellular carcinoma in adults. Mutations in other serpins leading to polymerization are implicated in diseases such as AT deficiency [Picard et al., 2003], C1-inhibitor deficiency [Eldering et al., 1995] and familial dementia associated with Collins bodies [Davis et al, 1999]. The structure of the α_1-PI polymer and in general the structure of a serpin polymer has been a subject of investigation for over 15 years; nevertheless, it remains unknown.

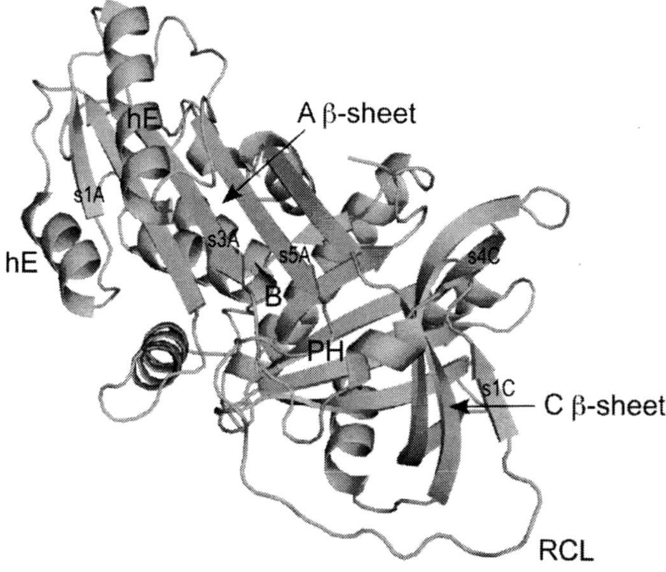

Figure 1. A ribbon diagram of α_1-PI (PDB:1QLP). The location of important elements of the structure including proximal hinge (PH) and breach (B) is indicated. Drawn using PyMOL[1].

Established Hypotheses

The structure of a serpin consists of three β-sheets (A-C), nine helices (A-I), and some interconnecting loops. A very important feature of this structure, playing both a functional

[1] DeLano, W.L. The PyMOL Molecular Graphics System (2002) on World Wide Web http://www.pymol.org

and a structural role, is the RCL that is exposed on the surface and serves as bait in serpin – protease interactions. Observation that RCL can be inserted into the A β-sheet or can associate with the A and C β-sheets as an edge strand led to several loop-sheet models of the polymer. Although the loop-A sheet model [Lomas et al., 1992] has never been proven, it is still generally accepted that *in vivo* α_1-PI polymerizes through sequential insertion of the RCL of one molecule between the central strands of the A β-sheet of another (Figure 2)[2].

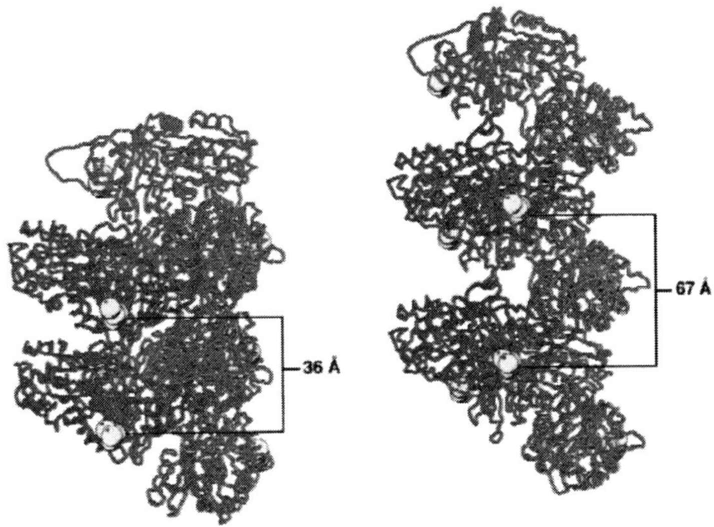

Figure 2. A schematic representation of an original open helical conformation of the loop-A sheet polymer with cysteine residues over 60 Å apart (right) and a modified closed helical conformation with cysteine residues close to each other and available for resonance energy transfer (left). The original model was modified to accommodate resonance energy transfer observations. Reprinted from Mahadeva et al., 1999 with permission of the American Society for Clinical Investigation. Copyright © American Society for Clinical Investigation. All rights reserved.

The loop-C sheet and strand s7A models, which were based on interactions seen in the crystals of other serpins, AT [Carrell et al., 1994] and PAI-1 [Sharp et al., 1999], also assumed that the mechanism of polymerization involves interaction of the RCL with a β-sheet. However, according to these models, the RCL of one molecule associates as an additional (edge) strand with the C or A sheet of another molecule, respectively. Recently, a new loop C-sheet mode of interactions was observed between serpin molecules in the crystal structure of a bacterial serpin, tengpin (tengpinΔ42) [Zhang et al., 2008], and this interaction was used as a basis for a loop-C sheet model of polymerization of latent serpins.

To explain the differences seen during biophysical characterization of polymers formed under different conditions, it has been speculated that different polymerization mechanisms operate depending on the variant and polymerization conditions. For example the loop-A

[2] After this article was accepted for publication two additional models of serpin polymers were proposed [Yamasaki et al., 2008, Tsutsui et al., 2008]. Publication of one of these models appeared to immediately lessen the acceptance of the loop A-sheet model [Whisstock and Bottomley, 2008]. Nevertheless, the significance of these models needs to be established.

sheet model was accepted for polymerization of the Z-variant [Lomas et al., 1992; Mallya et al.,2007] and the loop-C sheet model was proposed for AT [Carrell et al., 1994], the Mmalton variant polymer [Lomas et al., 1995], and both models concurrently were proposed for polymer formed in guanidine [Koloczek et al., 1996].

FACTS

It is unclear whether the loop-sheet interactions visible in crystals, which were the basis for the loop-sheet models, have a physiological relevance. They appear to be non-specific and may demonstrate the propensity of the RCL for interactions with any β-sheet. These interactions are not stable since upon crystal dissolution the protein dissociates to monomers [Carrell et al., 1994, Sharp et al., 1999, Zhang et al., 2008]. Thus, it appears that such interactions cannot be responsible for formation of stable polymers from the Z-variant *in vivo* or normal α_1-PI under mild denaturating conditions *in vitro*.

The widely accepted loop A-sheet model of the α_1-PI polymer was based on the observation that polymerization can be blocked in the presence of the RCL peptides [Lomas et al., 1992]. Earlier, a similar hypothesis was put forward based on the observation that α_1-PI forms complexes with RCL peptides, which integrate into its A β-sheet [Schulze et al, 1990]. The mobility of the RCL and its ability of insertion into the A β-sheet was demonstrated when the first crystal structure of α_1-PI was solved [Loebermann et al., 1984]. Currently, it is well established that serpins can insert their intact RCL between the strands 3 and 5 of their A β-sheet and form the latent form in *vivo* and *in vitro*. It is also known that serpins insert their cleaved RCL during protease inhibition into their A β-sheet and they can insert peptides derived from the sequence of the RCL or computer-designed (for a review of serpin crystal structures see Marszal and Shrake, 2006). However, it is uncertain whether the ability to insert the loop into the A β-sheet has any relevance to the structure of the serpin polymer. The fact that the inserted forms in general do not polymerize (latent form of neuroserpin is an exception [Onda et al., 2005] demonstrates solely that these forms are more stable than their normal counterparts.

It appears that some experimental results may not be easily explained with the accepted loop-A sheet model. Dissociation of polymers, which takes place in the presence of peptides and begins with internal fragmentation of the chain [Chowdhury et al., 2007], suggests that the space between strands s3A and s5A is, in the polymers, unoccupied and capable to accept a peptide and that peptide binding induces a conformational change, which leads to polymer dissociation (unless the loop insertion is reversible and peptides insert when the loop dissociates from the A β-sheet). Also, mutants' association rates with peptides do not correlate with their propensity to polymerize [Lomas et al., 1995, Parfrey et al., 2003].

The loop-A sheet model also appears to be in conflict with EM images. On electron micrographs, the polymers have the appearance of "beads on a string" (Figure 3). Regardless of whether they were formed from the Z-variant in the liver [Lomas et al., 1992] or from the S_{iiyama} [Lomas et al., 1993] or M_{malton} [Lomas et al., 1995] variants and were isolated from plasma, or whether they were formed *in vitro* from the Z [Lomas et al., 1992] or I variant [Mahadeva et al., 1999] or from normal α_1-PI exposed to the elevated temperature [Lomas et al., 1995, Stein and Carrell, 1995], low pH [Devlin et al., 2002] or proteolitically cleaved α_1-

PI [Mast et al., 1992], the polymer chains have a similar linear and relaxed form; some of them form circlets [Lomas et al., 1993, Mast et al., 1992]. This appearance, which is consistent with the structure of polymers formed from cleaved α_1-PI [Huntington et al., 1999; Dunstone et al., 2000], contrasts with the compactness of the loop-A sheet model proposed for the polymers formed from the intact protein (Figure 2 and Figure 3).

Furthermore, resonance energy transfer experiments using α_1-PI variants having the single cysteine, C232, labeled with fluorescent probes showed that the distance between cysteines in the polymer is shorter than 50 Å and possibly even shorter than 35 Å [Mahadeva et al., 1999]. This was in contrast with over 60 Å distance deduced from the loop-A sheet model. Thus, the model was modified to accommodate the resonance energy transfer observation (Figure 2) and became even more compact and, therefore, even more distant from what is seen by EM.

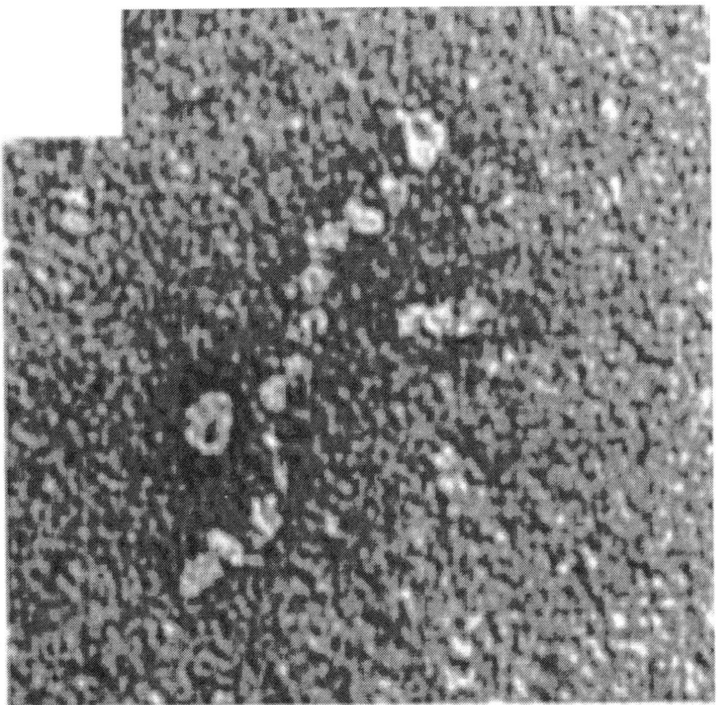

Figure 3. Electron micrograph of Z α_1-PI polymer viewed at 220,000 magnification. Reprinted from Carrell and Lomas, 2002 with permission of the Massachusetts Medical Society.

Although limited data relate directly to the structure of the serpin polymers, there is a wealth of information regarding conformational changes that occur within the serpin molecule during polymerization. These changes involve the release of the RCL and conformational changes in the vicinity of a highly conserved tryptophan residue (W194 in α_1-PI) and in the F-helix.

The release of the RCL during polymerization was demonstrated by the fact that crosslinking α_1-PI strands s1C and s2C with a disulfide bond prevented polymerization [Chang et al., 1997]. This was consistent with a polymerogenic nature of the most clinically relevant E342K mutation of α_1-PI in the proximal hinge and mutations engineered in the distal hinge region (positions P6'-P8') of the RCL, which likely destabilize the loop [Bottomley et al., 2001].

Conformational changes in the A β-sheet were suggested by changes in intrinsic fluorescence of W194, which is located directly under the proximal hinge of the RCL and close to the breach region of the A β-sheet [Dafforn et al., 1999, Tew and Bottomley, 2001]. Partial disruption of the A β-sheet during formation of a polymerogenic form can be also implied by the observation that insertion of a short peptide (6-mer) corresponding to the P14-P9 segment of the RCL induced polymerization of AT [Chang et al., 1997].

Structural changes also involve helix F and loop connecting the helix to the strand s3A (thFs3A) in the A β-sheet. This region was considered to play a significant role in preventing loop-sheet polymerization [Bruce et al. 1994]. Recently, it has been shown that crosslinking loop thFs3A and strand s3A in α_1-PI with an engineered disulfide bond significantly retarded polymerization [Baek et al., 2007], which suggests that the detachment of the loop from strand s3A occurs during the polymerization process. Accelerated polymerization of the I157L and D149A mutants also suggests that the interaction between helix F and the A β-sheet and interactions within helix F are modified during polymerization [Cabrita et al. 2004].

Interestingly, the structure of plasminogen activator inhibitor-2 (PAI-2) is locked in the polymerogenic state by an intrinsic disulfide bond, which upon reduction abrogates polymerization [Wilczynska et al., 2003]. The disulfide bond is formed between C161, located at the N-terminus of helix F, and C79, which is in the middle of the CD-loop, a loop present only in a few ov-serpins. The structure of the polymerogenic form of PAI-2, unfortunately, appears to be unknown; the structures available in the Protein Data Bank were solved for loop 66-98 deletion mutants.

An Alternative Model

The proximity of cysteines [Mahadeva et al., 1999] and cysteine-attached fluorophore protection from solvent during polymerization [Dafforn et al., 1999] suggests that α_1-PI monomers may associate in the head-to-head fashion during polymerization and that the cysteine area is involved in the intermolecular interface. This is consistent with the polymerization mode we observed for the disulfide-linked dimer of α_1-PI, which, in buffer under physiological conditions, spontaneously forms linear polymers that have EM appearance similar to all other α_1-PI polymers formed *in vivo* and *in vitro* [Marszal et al., 2003]. We proposed that in the polymers not containing disulfide bonds, the molecules can be oriented in the same manner. However, except for the requirement of the RCL release from the C β-sheet, the more detailed information regarding polymerogenic conformation of α_1-PI and possible head-to-head interactions in the vicinity of C232 between α_1-PI molecules is lacking. Since this area is rich in the β-sheet structure, a number of β-strand interactions seem possible, including association of the RCL as strand s5C and formation of an 8-strand β-barrel during head-to-head association of the monomers.

Polymerization of the disulfide-linked dimer demonstrated that α_1-PI is capable of forming tail-to-tail interactions at the site of the molecule opposite to the location of the RCL.

Interestingly, that part of the molecule is implicated in protein-protein interactions and is naturally used to stabilize serpin structure against conformational changes and transformation into the latent form [Zhou et al., 2003, Zhang et al., 2007]. It appears to have an ideal structure for potential interactions between two α_1-PI molecules, the association of which may occur through an antiparallel connection of two s1A strands, leading to the formation of a 10-strand β-sheet extending across two serpin molecules (Figure 4). In the case of the normal serpin conformation, polymerization appears to be prevented by a twist of strand s1A from the plane of the A β-sheet, which is one of the common strategies used to suppress edge-to-edge β-sheet interactions [Richardson and Richardson, 2002]. However, during conformational changes involving helix F that occur during polymerization, strand s1A may be brought into the β-sheet plane and allow for its extension.

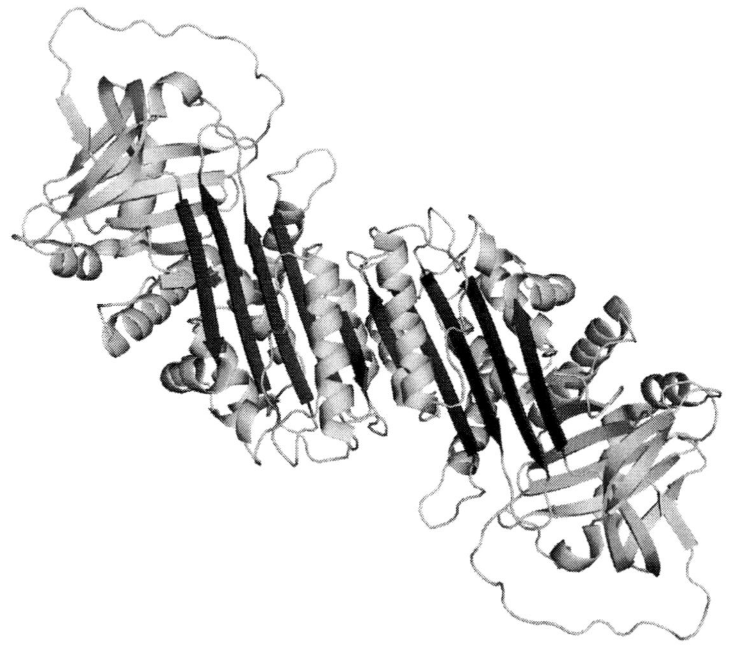

Figure 4. Model of the tail-to-tail association in the head-to-head/tail-to-tail polymer. The 10-strand β-sheet is shaded dark grey. Drawn using PyMOL.

This hypothesis can be supported with several observations. Relatively high flexibility in the area, which includes helix E, strand s1A, and helix F, is observed in serpin crystal structures (e.g., PDB:1QLP, 1WZ9, 1YXA) and in the α_1-PI structure in solution [Tsutsui et al., 2006] and, in addition, this area becomes protected from hydrogen-deuterium exchange in the polymer [Tsutsui, 2008]. Conformational changes during polymerization within helix F and loop thFs3A have been demonstrated [Baek et al., 2007]. In addition, the helix E-s1A-helix F region is involved in some interesting natural intra- and intermolecular interactions, e.g., in the case of tengpin, with its N-terminus, which protects this serpin from conformational changes and polymerization [Zhang et al., 2007] and, in the case of PAI-1, with an allosteric regulator, vitronectin, which protects PAI-1 from transformation to the latent form [Zhou et al, 2003]. The conformational switch between polymerogenic and nonpolymerogenic form regulated by the formation of a disulfide bond in PAI-2 is also located in this area [Wilczynska et al., 2003]. These observations suggest that there is a direct link between conformational changes occurring in the RCL area, in the serpin tail, and likely in the A β-sheet. Crystal structures of normal serpins (e.g., PDB:1QLP) and inserted serpins (e.g., PDB:1IZ2) show that helix F and loop thFs3A before and after incorporation of the RCL interact with strands s2A, s3A and s5A, protecting the integrity of the A β-sheet and stabilizing nonpolymerogenic conformation. When the RCL is not inserted, helix F may acquire a conformation that brings strand s1A to the A β-sheet plane and allows association with another serpin molecule through strand s1A-s1A interactions.

Recently, it has been demonstrated that α_1-PI polymerization can be blocked with small molecules targeting a pocket, which was thought to be distinct from the interface involved in polymerization [Mallya et al., 2007]. According to the head-to-head/tail-to-tail model, this area may be directly involved in conformational changes occurring in helix E and strand s1A during polymerization. The pocket is located next to helix E and strand s1A, and H139 which resides at the pocket entrance is situated in the loop connecting strand s1A and helix E.

While understanding the structure of α_1-PI is necessary for rational drug design, accumulating evidence seems to support the notion that currently established models of serpin polymers do not explain the experimental data well. It is clear that continued research is necessary. As increasing evidence accumulates, new hypotheses are being put forward. The title of a presentation made at the recent serpin conference Serpins 2008 suggests that a new model of the polymer is being proposed based on the crystal structure of a domain-swapped dimer [Yamasaki, 2008][3]. A new version of the loop-A sheet model has also been recently suggested based on hydrogen-deuterium exchange mass spectrometry analysis [Tsutsui, 2008][4]. Since the new models, including the model proposed herein, are still based on a limited amount of information, it is uncertain how closely they reflect the actual structure of the polymer. Nevertheless, these new models, and observations used to support them, may stimulate further research to explain experimental data which cannot be explained with the current models.

REFERENCES

Baek, JH; Im, H; Kang, UB; Seong, KM; Lee, C; Kim, J; Yu, MH. Probing the local conformational change of α_1-antitrypsin. *Protein Sci.*, 2007 16, 1842-50.

Bottomley, SP; Lawrenson, ID; Tew, D; Dai, W; Whisstock, JC; Pike, RN. The role of strand 1 of the C β-sheet in the structure and function of α_1-antitrypsin. *Protein Sci.*, 2001 10, 2518-24.

Bruce, D; Perry, DJ; Borg, JY; Carrell, RW; Wardell, MR. Thromboembolic disease due to thermolabile conformational changes of antithrombin Rouen-VI (187 Asn→Asp). *J. Clin. Invest.*, 1994 94, 2265-74.

[3] Published after acceptance of this article for publication [Yamasaki et al., 2008]

[4] Published after acceptance of this article for publication [Tsutsui et al., 2008]

Cabrita, LD; Dai, W; Bottomley, SP. Different conformational changes within the F-helix occur during serpin folding, polymerization, and proteinase inhibition. *Biochemistry*, 2004 43, 9834-9.

Carrell, RW; Stein, PE; Fermi, G; Wardell, MR. Biological implications of a 3 Å structure of dimeric antithrombin. *Structure*, 1994 2, 257–270.

Carrell, RW; Lomas, DA. Alpha$_1$-antitrypsin deficiency – a model for conformational diseases. *N. Engl. J. Med.*, 2002 346, 45-53.

Chang, WS; Whisstock, J; Hopkins, PCR; Lesk, AM; Carrell, RW; Wardell, MR. Importance of the release of strand 1C to the polymerization mechanism of inhibitory serpins. *Prot. Sci.*, 1997 6, 89-98.

Cho, YL; Chae, YK; Jung, CH; Kim, MJ; Na, YR; Kim, YH; Kang, SJ; Im, H. (2005) The native metastability and misfolding of serine protease inhibitors. *Protein Pept. Lett.*, 2005 12, 477-81.

Chowdhury, P; Wang, W; Lavender, S; Bunagan, MR; Klemke, JW; Tang, J; Saven, JG; Cooperman, BS; and Gai, F. Fluorescence correlation spectroscopic study of serpin depolymerization by computationally designed peptides *J. Mol. Biol.*, 2007 369, 462-473.

Dafforn, TR; Mahadeva, R; Elliott, PR; Sivasothy, P; Lomas, DA. A kinetic mechanism for the polymerization of α_1-antitrypsin. *J. Biol. Chem.*, 1999 274, 9548-55.

Davis, RL; Shrimpton, AE; Holohan, PD; Bradshaw, C; Feiglin, D; Collins, GH; Sonderegger, P; Kinter, J; Becker, LM; Lacbawan, F; Krasnewich, D; Muenke, M; Lawrence, DA; Yerby, MS; Shaw, CM; Gooptu, B; Elliott, PR; Finch, JT; Carrell, RW; Lomas DA. Familial dementia caused by polymerization of mutant neuroserpin *Nature*, 1999 401, 376-9.

Devlin, GL; Chow, MKM; Howlett, GJ; Bottomley, SP. Acid denaturation of α_1-antitrypsin: Characterization of a novel mechanism of serpin polymerization. *J. Mol. Biol.*, 2002 324; 859-870.

Dunstone, MA; Dai, W; Whisstock, JC; Rossjohn, J; Pike, RN; Feil, SC; Le Bonniec, BF; Parker, MW; Bottomley, SP. Cleaved antitrypsin polymers at atomic resolution. *Protein Sci.*, 2000 9, 417-20.

Eldering, E; Verpy, E; Roem, D; Meo, T; Tosi, M. COOH-terminal substitutions in the serpin C1 inhibitor that cause loop overinsertion and subsequent multimerization. *J. Biol. Chem.*, 1995 270, 2579-87.

Huntington, JA; Pannu, NS; Hazes, B; Read, RJ; Lomas, DA; Carrell, RW. A 2.6 Å structure of a serpin polymer and implications for conformational disease. *J. Mol. Biol.*, 1999 293, 449-55.

Irving, JA; Pike, RN; Lesk, AM; Whisstock, JC. Phylogeny of the serpin superfamily: Implications of patterns of amino acid conservation for structure and function. *Genome Res.*, 2000 10, 1845-64.

Koloczek, H; Guz, A; Kaszycki, P. Fluorescence-detected polymerization kinetics of human α_1-antitrypsin. *J. Protein Chem.*, 1996 15, 447-54.

Loebermann; H; Tokuoka, R; Deisenhofer, J.; Huber, R. Human α_1-proteinase inhibitor. Crystal structure analysis of two crystal modifications, molecular model and preliminary analysis of the implications for function. *J. Mol. Biol.*, 1984 177, 531-56.

Lomas, DA; Evans, DL; Finch, JT; Carrell, RW. The mechanism of Z α_1-antitrypsin accumulation in the liver. *Nature*, 1992 357, 605-7.

Lomas, DA; Finch, JT; Seyama, K; Nukiwa, T; Carrell, RW. α_1-Antitrypsin S_{iiyama} (Ser53→Phe). Further evidence for intracellular loop-sheet polymerization. *J. Biol. Chem.*, 1993 268, 15333-5.

Lomas, DA; Elliott, PR; Sidhar, SK; Foreman, RC; Finch, JT; Cox, DW; Whisstock, JC; Carrell, RW. α_1-antitrypsin Mmalton (Phe^{52}-deleted) forms loop-sheet polymers *in vivo*. Evidence for the C sheet mechanism of polymerization. *J. Biol. Chem.*, 1995 270, 16864–70.

Mahadeva, R; Chang, W-SW; Dafforn, TR; Oakley, DJ; Foreman, RC; Calvin, J; Wight, DGD; Lomas, DA. Heteropolymerization of S, I, and Z α_1-antitrypsin and liver cirrhosis. *J. Clin. Invest.*, 1999 103, 999–1006.

Mallya, M; Phillips, RL; Saldanha, SA; Gooptu, B; Brown, SC; Termine, DJ; Shirvani, AM; Wu, Y; Sifers, RN; Abagyan, R; Lomas DA. Small molecules block the polymerization of Z α_1-antitrypsin and increase the clearance of intracellular aggregates. *J. Med. Chem.*, 2007 50, 5357-63.

Marszal, E; Danino, D; Shrake, A. A novel mode of polymerization of α_1-proteinase inhibitor. *J. Biol. Chem.*, 2003 278, 19611-8.

Marszal, E; Shrake, A. Serpin crystal structure and serpin polymer structure. *Arch. Biochem. Biophys.*, 2006 453, 123-9.

Mast, AE; Enghild, JJ; Salvesen, G. Conformation of the reactive site loop of α_1-proteinase inhibitor probed by limited proteolysis. *Biochemistry*, 1992 31, 2720-8.

Onda, M; Belorgey, D; Sharp, LK; Lomas, DA. Latent S49P neuroserpin forms polymers in the dementia familial encephalopathy with neuroserpin inclusion bodies. *J. Biol. Chem.*, 2005 280, 13735-41.

Parfrey, H; Mahadeva, R; Ravenhill, NA; Zhou, A; Dafforn, TR; Foreman, RC; Lomas, DA. Targeting a surface cavity of α_1-antitrypsin to prevent conformational disease. *J. Biol. Chem.*, 2003 278, 33060-6.

Picard, V; Dautzenberg, MD; Villoutreix, BO; Orliaguet, G; Alhenc-Gelas, M; Aiach, M. Antithrombin Phe229Leu: a new homozygous variant leading to spontaneous antithrombin polymerization in vivo associated with severe childhood thrombosis. *Blood*, 2003 102, 919-25.

Richardson, JS; Richardson, DC. Natural β-sheet proteins use negative design to avoid edge-to-edge aggregation. *Proc. Natl. Acad. Sci. USA*, 2002 99, 2754-9.

Schulze, AJ; Baumann, U; Knof, S; Jaeger, E; Huber, R; Laurell, C-B. Structural transition of α_1-antitrypsin by a peptide sequentially similar to β-strand s4A. *Eur. J. Biochem.*, 1990 194, 51–6.

Sharp, AM; Stein, PE; Pannu, NS; Carrell, RW; Berkenpas, MB; Ginsburg, D; Lawrence, DA; Read, RJ. The active conformation of plasminogen activator inhibitor 1, a target for drugs to control fibrinolysis and cell adhesion. *Structure Fold. Des.*, 1999 7, 111-8.

Stein, PE; Carrell, RW. What do dysfunctional serpins tell us about molecular mobility and disease? *Nat. Struct. Biol.*, 1995 2, 96-113.

Tew, D; Bottomley, SP. Probing the equilibrium denaturation of the serpin α_1-antitrypsin with single tryptophan mutants; Evidence for structure in the urea unfolded state. *J. Mol. Biol.*, 2001 313, 1161-9.

Tsutsui, Y; Liu, L; Gershenson, A; Wintrode, PL. The conformational dynamics of a metastable serpin studied by hydrogen exchange and mass spectrometry *Biochemistry*, 2006 45, 6561-9.

Tsutsui, Y. Exploring Functional and Folding Energy Landscapes by Hydrogen-Deuterium Exchange Mass Spectrometry. Ph.D. Dissertation 2008. Case Western Reserve University / OhioLINK. Available from: URL: http://www.ohiolink.edu/etd/send-pdf.cgi/Tsutsui%20Yuko.pdf?acc_num=case1196199391. Accessed July 27, 2008.

Tsutsui, Y; Kuri, B; Sengupta, T; Wintrode, PL. The structural basis of serpin polymerization studied by hydrogen/deuterium exchange and mass spectrometry. *J. Biol. Chem.*, 2008 283, 30804-30811.

Whisstock, JC; Bottomley, SP. Serpins' mystery solved. Nature, 2008 455, 1189-1190.

Wilczynska, M; Lobov, S; Ohlsson, PI; Ny, T. A redox-sensitive loop regulates plasminogen activator inhibitor type 2 (PAI-2) polymerization. *EMBO J.*, 2003 22, 1753-61.

Zhang, Q; Buckle, AM; Law, RHP; Pearce, MC; Cabrita, LD; Lloyd, GJ; Irving, JA; Smith, AI; Ruzyla, K; Rossjohn, J; Bottomley, SP; Whisstock, JC. The N terminus of the serpin, tengpin, functions to trap the metastable native state. *EMBO Rep.*, 2007 8, 658-63.

Zhang, Q; Law, RHP; Bottomley, SP; Whisstock, JC; Buckle, AM. A structural basis for loop C-sheet polymerization in serpins. *J. Mol. Biol.*, 2008 376, 1348-1359.

Zhou, A; Huntington, JA; Pannu, NS; Carrell, RW; Read, RJ. How vitronectin binds PAI-1 to modulate fibrinolysis and cell migration. *Nat. Struct. Biol.*, 2003 10, 541-4.

Yamasaki, M. Crystal structure of a domain-swapped dimer reveals the molecular basis of serpin polymerization. *Serpins* 2008, Leuven, Belgium, July 12-16, 2008.

Yamasaki, M; Li, W; Johnson, DJ; Huntington, JA. Crystal structure of a stable dimer reveals the molecular basis of serpin polymerization. Nature, 2008 455, 1255-1258.

Ligands, Polymers and Amino Acids
Editor: Tsisana Shartava
ISBN: 978-1-61122-793-2

INCREASE IN AMINO ACIDS AND AMMONIA DURING LIQUEFACTION OF SEMEN

*Thomas W. Stief**
Department of Clinical Chemistry,
University Hospital Giessen and Marburg, Germany

ABSTRACT

Fresh normal ejaculate clots by action of about 0.05 IU/ml thrombin on a fibrinogen-like compound, and subsequently lyses by action of about 0.1 IU/ml plasmin and about 300 % (of blood plasma norm) carboxypeptidase. Carboxypeptidases sequentially liberate amino acids, that were quantified in the present work: the most abundant amino acids in fresh seminal plasma are serine (6.09 mM), threonine (4.61 mM), lysine (3.45 mM), tyrosine (3.18 mM), glycine (2.79 mM), and arginine (2.29 mM). The concentration of the amino acids approximately doubles within 45 min (37°C), with an exceptional 12fold increase of glutamic acid. When compared to blood plasma, fresh seminal plasma contains 226fold gamma-glutamyl-transferase (γ-GT), 40fold ammonia, 21fold creatinine, and 18fold glutamate dehydrogenase (GLDH). Ammonia (NH_3), generated by glutamate dehydrogenase (GLDH), together with arginine or creatinine might protect sperm cells from female defence cells. Ammonia and natural guanidine derivatives might be physiologic immunsuppressors.

Keywords: Semen, liquefaction, GLDH, γ-GT, amino acids, glutamic acid, ammonia.

INTRODUCTION

Knowledge of the exact composition of human seminal plasma, which is rather unstable due to the presence of proteases, is of importance to understand fertility [1-8]. Recently, high activities of carboxypeptidase in seminal plasma have been described [9-11]. Carboxypeptidases are proteolytic enzymes that cleave peptide bonds at the carboxy terminus of proteins, thereby sequentially liberating amino acids and thus degrading proteins.

* Correspondence to: PD Dr. med. T. Stief, Department of Clinical Chemistry, University Hospital, D-35043 Marburg, Germany, Email: thstief@med.uni-marburg.de; Tel. : +49-6421-28 64471; FAX: +49-6421-28 65594

Physiologically, the fresh ejaculate clots by about 20 ng/ml tissue factor - mediated generation of about 0.05 IU/ml thrombin that acts on about 20-30 g/l of a fibrinogen-like compound (assayed antigenically [12]), generating a coagulum similar to fibrin [13,14]. The seminal clot lyses by action of about 0.1 IU/ml plasmin [3,15,16,80]: during liquefaction of the coagulum, its fibers are transformed into globular bodies that form the homogeneous liquefied semen [13], wherein the spermatozoa can freely move. Seminal clotting and sperm motility correlate negatively [17]. The aim of the present work was the clinical chemical analysis of seminal plasma immediately after ejaculation (S_0) and after a liquefaction time of 45 min (37°C) (S_{45}).

Material and Methods

The experiment was performed 3 times with 3 ml freshly ejaculated normal semen after 3 days of abstinence, 1.5 ml was centrifuged (3000 g for 5 min at 23°C) immediately, and 1.5 ml was centrifuged after 45 min (37°C). The samples were diluted 1+4 with H_2O immediately before duplicate clinical-chemical analysis, which was performed within 1 h (23°C). Amino acids were determined by a cation exchange amino acid analyzer (LC 3000; Eppendorf-Biotronic, Hamburg, Germany). The concentrations found were compared with the respective mean values of normal plasma or with the upper normal value of plasma. The clinical chemical parameters were determined by a Hitachi 917 analyzer (Roche, Basel, Switzerland). The 45 min-results were compared with the basal controls, and were tested for significance ($p< 0.05$) using the X^{2X} test [12]. The mean values were calculated, the standard deviations of all analyte-determinations were < 10 %.

Sodium dodecylsulfate gel electrophoresis (SDS-PAGE): samples of fresh seminal plasma without liquefaction time (sample 1) and with 45 min (37°C) liquefaction time (sample 2) were pre-diluted 1:20 with 0.9 % NaCl, pooled frozen/thawed normal citrated blood plasma was pre-diluted 1:20 and 1:50 with 0.9 % NaCl. These samples were further diluted 1:2 with sample buffer consisting of 10 mM Tris, 2 % SDS, 0.01 % bromphenol blue, 1 mM EDTA, pH 8, and run for 1 h in a PhastSystem with pre-defined molecular weight markers (Amersham Pharmacia Biotech, Freiburg, Germany), using PhastGel Gradient 8-25 (6-300 kilo Dalton (kDa)) and PhastGel buffer strips (0.2 M tricine, 0.2 M Tris, 0.55 % SDS, pH 8.1). Then the SDS-PAGE gels were silver stained for 6.5 min with 0.4 % silver nitrate.

Results and Discussion

Table 1 demonstrates that all amino acids occur in seminal plasma in an up to about 50fold higher concentration when compared to blood plasma. The exception is methionine that was not found in seminal plasma; methionine in seminal plasma might be metabolized or consumed immediately [18]. The most abundant amino acids in fresh seminal plasma are serine (6.09 mM), threonine (4.61 mM), lysine (3.45 mM), tyrosine (3.18 mM), glycine (2.79 mM), and arginine (2.29 mM). Serine, threonine, and tyrosine have already been described as being the principal amino acids found in seminal plasma [19]. The most abundant amino acids in 45 min (37°C) incubated seminal plasma are serine (9.01 mM), glutamine (6.86

mM), glutamic acid (6.38 mM), tyrosine (5.69 mM), threonine (5.37 mM), glycine (5.35 mM), leucine (4.28 mM), arginine (4.21 mM), histidine (4.19 mM), and lysine (4.1 mM). Overall the concentration of the amino acids approximately doubles within 45 min (37°C), with an exceptional 12fold increase of glutamic acid.

Table 1. Amino acid concentrations in liquefying semen

Amino Acid [mM]	Plasma	$Semen_{0min}$	Ratio S_0/P	$Semen_{45min}$	RatioS_{45}/P	RatioS_{45}/S_0
Alanine	0.15-0.49	0.58	1.8	1.83	5.7	3.2
Arginine	0.03-0.11	2.29	33	4.21	60	1.8
Asparagine	0.03-0.07	1.07	21	3.07	61	2.9
Asp. acid	< 0.01	0.52	52	1.46	146	2.8
Cystin	0.03-0.05	0.09	2.3	0.12	3	1.3
Glutamine	0.34-0.74	1.45	2.7	6.86	13	4.7
Glut. acid	0.01-0.05	0.52	17	6.38	213	12.3
Glycine	0.10-0.38	2.79	12	5.35	22	1.9
Histidine	0.07-0.10	2.08	24	4.19	49	2.0
Isoleucine	0.03-0.08	1.41	26	2.64	48	1.9
Leucine	0.08-0.16	2.08	17	4.28	36	2.1
Lysine	0.11-0.21	3.45	22	4.10	26	1.2
Methionine	0.01-0.03	< 0.01	0	< 0.01	0	-
Ornithine	0.04-0.10	0.23	3.3	0.27	3.9	1.2
PhenAla.	0.04-0.06	0.80	16	1.62	32	2.0
Proline	0.10-0.24	0.60	4.7	1.12	6.6	1.9
Serine	0.07-0.18	6.09	49	9.01	72	1.5
Taurine	0.02-0.08	1.25	25	1.07	21	0.9
Threonine	0.08-0.20	4.61	33	5.37	38	1.2
Tryptophane	0.02-0.05	0.14	4.0	0.32	9.1	2.3
Tyrosine	0.03-0.08	3.18	58	5.69	103	1.8
Valine	0.15-0.25	1.35	6.8	3.52	18	2.6

Table 2 shows the concentrations of important clinical chemical parameters in seminal plasma as compared to blood plasma. Some analytes are severalfold more concentrated (*) in fresh seminal plasma when compared to blood plasma: γ-GT (226*), ammonia (40*), phosphate (29*), creatinine (21*), GLDH (18*), LDH_{1+2} (13*), GOT (10*), Mg (9*), LDH (8*), K^+ (7*), and Ca^{++} (3*), the latter being of importance in sperm motility [20]. Albumin, IgG, IgA, IgM, CRP, and transferrin could not be detected in seminal plasma. Pseudocholinesterase (PCHE) is found in seminal plasma at only at about 2 % of the blood plasma level. Within 45 min liquefaction time (37°C) ammonia increased from 2.1 mM to 6.8 mM and lactate from 2.2 mM to 3.3 mM; the activity of nearly all clinical chemical enzymes was then approximately 50 % of that before liquefaction, i.e. enzymes are proteolytically degraded and/or inactivated.

Figure 1 demonstrates that the protein band of a protein of about 50 kDa in fresh seminal plasma diminishes to some extent whereas the concentrations of high molecular weight proteins of about 200 kDa and low molecular weight proteins of about 10-30 kDa increase.

Table 2. Clinical chemical parameters in liquefying semen

Parameter	Dimension	Plasma	Semen0min	Ratio S0/P	Semen45min	RatioS45/S0
Na+	mM	135-145	130	0.93	115	0.88
K+	mM	3.6-4.8	28.3	6.7	24.8	5.9
Ca++	mM	2.1-2.6	8.0	3.4	7.0	3.0
Mg++	mM	0.7-1.05	7.5	8.6	7.0	8.0
Phosphate	mM	0.84-1.5	33.8	28.9	33.0	28.2
Cl-	mM	101-111	63	0.6	65	0.6
GPT=ALT	U/l	< 45	58	1.3	38	0.8
GOT=AST	U/l	< 35	335	9.6	195	5.6
γ-GT	U/l	< 55	12428	226	6398	116
GLDH	U/l	< 6.4	118	18	76	12
AP	U/l	< 129	463	3.6	215	1.7
LDH	U/l	< 248	1933	7.8	1145	4.6
LDH1+2	U/l	72-182	1699	13.4	977	7.7
CK	U/l	< 171	555	3.2	268	1.6
P-Amylase	U/l	8-53	25	0.8	18	0.6
Lipase	U/l	< 60	50	0.8	40	0.7
PCHE	U/l	5320-12920	143	0.016	125	0.014
Bilirubin	mg/dl	0.2-1.0	< 0.2	0	< 0.2	0
Glukose	mg/dl	70-110	35	0.4	18	0.2
Creatinine	mg/dl	0.67-1.17	19.7	21.4	19.0	20.7
Urea	mg/dl	20-50	55	1.6	58	1.7
Uric acid	mg/dl	3.4-7.0	13.3	2.6	14.0	2.7
Cholesterol	mg/dl	< 200	30	0.2	10	0.1
Triglyc.	mg/dl	< 150	< 10	0	< 10	0
Iron	µmol/l	11-30	3	0.2	4	0.2
Protein	g/l	66-83	35	0.5	35	0.5
Albumin	g/l	35-52	< 10	0	< 10	0
IgG	g/l	7-16	< 0.3	0	< 0.3	0
IgA	g/l	0.7-4.0	< 0.1	0	< 0.1	0
IgM	g/l	0.4-2.3	< 0.1	0	< 0.1	0
CRP	mg/l	< 5	< 5	0	< 5	0
Myoglobin	µg/l	9.6-67	30	0.8	0	0
Ferritin	µg/l	30-400	183	0.9	85	0.4
Transferrin	g/l	2.0-3.6	< 0.1	0	< 0.1	0
Lactate	mM	0.5-2.2	2.2	1.6	4.4	3.3
Ammonia	µM	< 53	2112	40	6765	128

This might be due to the action of seminal transglutaminase and carboxypeptidase, respectively.

Of particular interest is the very high activity of gamma-glutamyl-transferase (γ-GT) [21] and of glutamine/glutamic acid in seminal plasma. γ-GT is an important membrane protein, that shuttles amino acids into cells, whereby an extracellular amino acid is coupled to glutamic acid (out of glutathione) [22-28]. The endothelial membrane ectoenzyme γ-GT is a marker for the blood brain barrier [27]. Seminal carboxypeptidases generate high concentrations of amino acids that might be uptaken by the sperm cells.

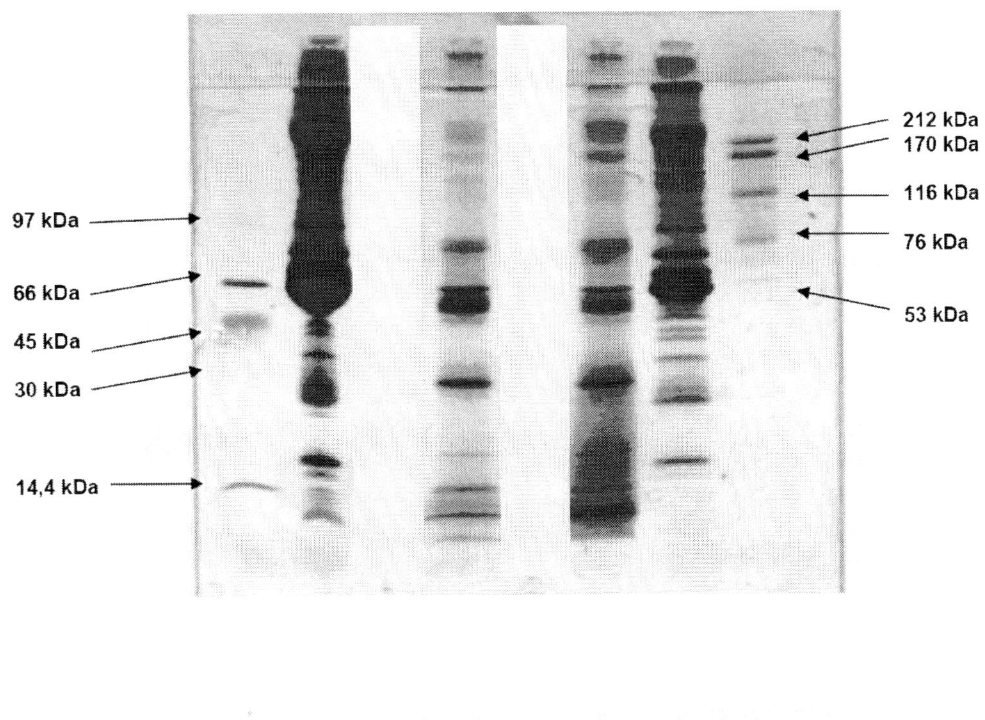

Figure 1. SDS-PAGE of seminal plasma. Seminal plasma without liquefaction time (sample 1) and with 45 min (37°C) liquefaction time (sample 2), both pre-diluted 1:20, and normal blood plasma, pre-diluted 1:20 (lane 2) and 1:50 (lane 5), were run in SDS-PAGE, followed by silver staining of the separated seminal plasma proteins and the low and high molecular weight standards (LMW and HMW; lanes 1 and 6).

These high concentrations of amino acids are favourable for the vitality and motility of spermatozoa [7,29-34] and the vitality of the oocyte [35-37]. Taurine seems to be important for sperm survival and sperm/oocyte fusion [37]. 5 mM proline, 10 mM glutamine, and 10-20 mM glycine stabilize spermatozoa [38]. Low liquefaction correlates to low amino-acids content [33].

The actions of γ-GT (induction of redox cycling [28]) or that of glutamate dehydrogenase (GLDH) [39] or the glutamine to glutamic acid decay [40] could generate the cell signals singlet oxygen ($^1O_2$) or ammonia (NH_3) [41]; ammonia at mM concentrations evokes oxidative/nitrosative stress, the mitochondrial permeability transition, and cell swelling [42-45]: it might be toxic for sperm attacking cells of the female immune response. The spermatozoa are also susceptible to high concentrations of reactive oxygen species (ROS) in their ambience [46-51], healthy spermatozoa tolerate some ROS concentrations [47]. Interestingly, spermatozoa via own NADPH-oxidase locally generate ROS, that has physiologic functions in spermatozoa motility and sperma/oocyte fusion [52-61]. NADPH oxidase is mainly located on the head domains of the spermatozoa [60], indicating the importance of oxidants in capacitation.

The ammonia- or arginine- mediated immunsuppression might be of importance for sperm cells to escape from the action of tissue monocytes, granulocytes, or lymphocytes [62-72]. Even high concentrations of the guanidine derivative creatinine do not impair sperm motility [73]. The biochemical composition of seminal plasma might be of interest to develop new physiologic immunsuppressors [74-79,81].

Acknowledgment

Britta Kosche is thanked for performing the miniaturized SDS-PAGE and the subsequent silver staining of the proteins bands on the gels.

References

[1] van der Merwe FH, Kruger TF, Oehninger SC, Lombard CJ. The use of semen parameters to identify the subfertile male in the general population. *Gynecol. Obstet. Invest.* 2005; 59: 86-91.

[2] Balerna M, Medici G, Mazzucchelli L, Bianda T, Marossi L, Colpi GM. Analytical biochemistry of seminal vesicle secretion: a challenge to andrological laboratories. *Andrologia* 1990; 22 Suppl 1: 166-77.

[3] Lwaleed BA, Goyal A, Delves GH, Cooper AJ. Seminal hemostatic factors: then and now. *Semin. Thromb. Hemost.* 2007; 33: 3-12.

[4] Hirsch I H, Jeyendran R S, Sedor J, Rosecrans R, Staas WE. Biochemical analysis of electroejaculates in spinal cord injured men: comparison to normal ejaculates. *Journal of Urology* 1991; 145: 73-77.

[5] Kaemmerer H, Mitzkat HJ. Ion-exchange chromatography of amino acids in ejaculates of diabetics. *Andrologia* 1985; 17: 485-7.

[6] Nissen HP, Schirren C, Kreysel HW, Heinze I. Amino acid composition of human semen from different clinical diagnoses. *Andrologia* 1979; 11: 109-12.

[7] Silvestroni L, Morisi G, Malandrino F, Frajese G. Free amino acids in semen: measurement and significance in normal and oligozoospermic men. *Arch. Androl.* 1979; 2: 257-61.

[8] Srivastava A, Chopra SK, Dasgupta PR. Biochemical analysis of human seminal plasma. II. Protein, non-protein nitrogen, urea, uric acid and creatine. *Andrologia* 1984; 16: 265-8.

[9] Lwaleed BA, Goyal A, Greenfield RS, Cooper AJ. Seminal thrombin-activatable fibrinolysis inhibitor: a regulator of liquefaction. *Blood Coagul Fibrinolysis* 2007; 18: 449-54.

[10] Willemse JL, Hendriks DF. A role for procarboxypepidase U (TAFI) in thrombosis. *Front Biosci.* 2007; 12: 1973-87.

[11] Willemse JL, Hendriks DF. Measurement of procarboxypeptidase U (TAFI) in human plasma: a laboratory challenge. *Clin. Chem.* 2006; 52: 30-6.

[12] Stief TW. The fibrinogen antigenic turbidimetric assay (FIATA): the X^{2X} test--the corrected chi-square comparison against the control-mean. *Clin. Appl. Thromb. Hemost.* 2007; 13: 73-100.

[13] Mandal A, Bhattacharyya AK. Some preliminary observations on the liquefaction of human semen. *Andrologia* 1985; 17: 228-33.

[14] de Lamirande E. Semenogelin, the main protein of the human semen coagulum, regulates sperm function. *Semin. Thromb. Hemost.* 2007; 33: 60-8.

[15] Van Dreden P, Audrey C, Aurélie R. Human seminal plasma fibrinolytic activity. *Semin. Thromb. Hemost.* 2007; 33: 21-8.

[16] Stief TW. Thrombin and plasmin activity in semen. *Blood Coagul Fibrinolysis* 2007; 18: 386-7.

[17] Ohta S, Wada H, Gabazza EC, Nobori T, Fuse H. Evaluation of tissue factor antigen level in human seminal plasma. *Urol. Res.* 2002; 30: 317-20.

[18] Laudat A, Lecourbe K, Guéchot J, Palluel AM. Values of sperm thiobarbituric acid-reactive substance in fertile men. *Clin. Chim. Acta* 2002; 325: 113-5.

[19] Fröhlich JU, Nissen HP, Heinze I, Schirren C, Kreysel HW. Free amino-acid composition of human seminal plasma in different andrological diagnoses. *Andrologia* 1980; 12: 162-6.

[20] Fakih H, MacLusky N, DeCherney A, Wallimann T, Huszar G. Enhancement of human sperm motility and velocity in vitro: effects of calcium and creatine phosphate. *Fertil Steril.* 1986; 46: 938-44.

[21] Chen F, Lu JC, Xu HR, Huang YF, Lu NQ. Preliminary investigations on the standardisation and quality control for the determination of γ-glutamyltranspeptidase activity in seminal plasma. *Andrologia* 2007; 39: 1-43.

[22] Okada T, Suzuki H, Wada K, Kumagai H, Fukuyama K. Crystal structure of the gamma-glutamyltranspeptidase precursor protein from Escherichia coli. Structural changes upon autocatalytic processing and implications for the maturation mechanism. *J. Biol. Chem.* 2007; 282: 2433-9.

[23] Kinlough CL, Poland PA, Bruns JB, Hughey RP. Gamma-glutamyltranspeptidase: disulfide bridges, propeptide cleavage, and activation in the endoplasmic reticulum. *Methods Enzymol.* 2005; 401: 426-49.

[24] Hawkins RA, O'Kane RL, Simpson IA, Viña JR. Structure of the blood-brain barrier and its role in the transport of amino acids. *J. Nutr.* 2006; 136: 218S-26S.

[25] Watanabe N, Dickinson DA, Liu RM, Forman HJ. Quinones and glutathione metabolism. *Methods Enzymol.* 2004; 378: 319-40.

[26] Pastore A, Federici G, Bertini E, Piemonte F. Analysis of glutathione: implication in redox and detoxification. *Clin. Chim. Acta* 2003; 333: 19-39.

[27] Zinke H, Möckel B, Frey A, Weiler-Güttler H, Meckelein B, Gassen HG. Blood-brain barrier: a molecular approach to its structural and functional characterization. *Prog. Brain Res.* 1992; 91: 103-16.

[28] Paolicchi A, Dominici S, Pieri L, Maellaro E, Pompella A. Glutathione catabolism as a signaling mechanism. *Biochem. Pharmacol.* 2002; 64: 1027-35.

[29] Brown-Woodman PD, White IG. Amino acid composition of semen and the secretions of the male reproductive tract. *Aust. J. Biol. Sci.* 1974; 27: 415-22.

[30] Van Thuan N, Wakayama S, Kishigami S, Wakayama T. New preservation method for mouse spermatozoa without freezing. *Biol. Reprod.* 2005; 72: 444-50.

[31] He S, Woods LC. Effects of glycine and alanine on short-term storage and cryopreservation of striped bass (Morone saxatilis) spermatozoa. *Cryobiology* 2003; 46: 17-25.

[32] Trimeche A, Yvon JM, Vidament M, Palmer E, Magistrini M. Effects of glutamine, proline, histidine and betaine on post-thaw motility of stallion spermatozoa. *Theriogenology* 1999; 52: 181-91.

[33] Hernvann A, Gonzales J, Troupel S, Galli A. Amino acid content of human semen in normal and infertility cases. *Andrologia* 1986; 18: 461-9.

[34] Keil M, Wetterauer U, Heite HJ. Glutamic acid concentration in human semen--its origin and significance. *Andrologia* 1979; 11: 385-91.

[35] Takahashi Y, Kanagawa H. Effects of glutamine, glycine and taurine on the development of in vitro fertilized bovine zygotes in a chemically defined medium. *J. Vet. Med. Sci.* 1998; 60: 433-7.

[36] Ka HH, Sawai K, Wang WH, Im KS, Niwa K. Amino acids in maturation medium and presence of cumulus cells at fertilization promote male pronuclear formation in porcine oocytes matured and penetrated in vitro. *Biol. Reprod.* 1997; 57: 1478-83.

[37] Guérin P, Ménézo Y. Hypotaurine and taurine in gamete and embryo environments: de novo synthesis via the cysteine sulfinic acid pathway in oviduct cells. *Zygote* 1995; 3: 333-43.

[38] Li Y, Si W, Zhang X, Dinnyes A, Ji W. Effect of amino acids on cryopreservation of cynomolgus monkey (Macaca fascicularis) sperm. *Am. J. Primatol.* 2003; 59: 159-65.

[39] Huijmans JG, Duran M, de Klerk JB, Rovers MJ, Scholte HR. Functional hyperactivity of hepatic glutamate dehydrogenase as a cause of the hyperinsulinism/ hyperammonemiasyndrome: effect of treatment. *Pediatrics* 2000; 106: 596-600.

[40] Murphy S, Munoz S, Parry-Billings M, Newsholme E. Amino acid metabolism during platelet storage for transfusion. *Br. J. Haematol.* 1992; 81: 585-90.

[41] Häussinger D, Schliess F. Glutamine metabolism and signaling in the liver. *Front Biosci.* 2007; 12: 371-91.

[42] Albrecht J, Norenberg MD. Glutamine: a Trojan horse in ammonia neurotoxicity. *Hepatology* 2006; 44: 788-94.

[43] Rose C. Effect of ammonia on astrocytic glutamate uptake/release mechanisms. *J. Neurochem.* 2006; 97 Suppl 1: 11-5.

[44] Norenberg MD, Rao KV, Jayakumar AR. Mechanisms of ammonia-induced astrocyte swelling. *Metab. Brain Dis.* 2005; 20: 303-18.

[45] Ziegler HK, Unanue ER. Decrease in macrophage antigen catabolism caused by ammonia and chloroquine is associated with inhibition of antigen presentation to T cells. *Proc. Natl. Acad. Sci. USA* 1982; 79: 175-8.

[46] Klebanoff S J, Smith DC. The source of H_2O_2 for the uterine fluid-mediated sperm-inhibitory system . *Biology of Reproduction* 1970; 3: 236-242.

[47] Fraczek M, Sanocka D, Kurpisz M. Interaction between leucocytes and human spermatozoa influencing reactive oxygen intermediates release. *Int. J. Androl.* 2004; 27: 69-75.

[48] Szczesniak-Fabianczyk B, Bochenek M, Smorag Z, Ryszka F. Effect of antioxidants added to boar semen extender on the semen survival time and sperm chromatin structure. *Reprod. Biol.* 2003; 3: 81-7.

[49] Baumber J, Vo A, Sabeur K, Ball BA. Generation of reactive oxygen species by equine neutrophils and their effect on motility of equine spermatozoa. *Theriogenology* 2002; 57: 1025-33.

[50] Erkkila K, Hirvonen V, Wuokko E, Parvinen M, Dunkel L. N-acetyl-L-cysteine inhibits apoptosis in human male germ cells in vitro. *J. Clin. Endocrinol. Metab.* 1998; 83: 2523-31.

[51] Zalata AA, Christophe AB, Depuydt CE, Schoonjans F, Comhaire FH. White blood cells cause oxidative damage to the fatty acid composition of phospholipids of human spermatozoa. *Int. J. Androl.* 1998; 21: 154-62.

[52] Alabi NS, Whanger PD, Wu AS. Interactive effects of organic and inorganic selenium with cadmium and mercury on spermatozoal oxygen consumption and motility in vitro. *Biol. Reprod.* 1985; 33: 911-9.

[53] de Lamirande E, Yoshida K, Yoshiike TM, Iwamoto T, Gagnon C. Semenogelin, the main protein of semen coagulum, inhibits human sperm capacitation by interfering with the superoxide anion generated during this process. *J. Androl.* 2001; 22: 672-9.

[54] Ball BA, Vo AT, Baumber J. Generation of reactive oxygen species by equine spermatozoa. *Am. J. Vet. Res.* 2001; 62: 508-15.

[55] Griveau JF, Le Lannou D. Reactive oxygen species and human spermatozoa: physiology and pathology. *Int. J. Androl.* 1997; 20: 61-9.

[56] de Lamirande E, Gagnon C. Human sperm hyperactivation and capacitation as parts of an oxidative process. *Free Radic. Biol. Med.* 1993; 14: 157-66.

[57] Babior BM. NADPH oxidase. *Curr. Opin. Immunol.* 2004; 16: 42-7.

[58] Roy SC, Atreja SK. Production of superoxide anion and hydrogen peroxide by capacitating buffalo (Bubalus bubalis) spermatozoa. Anim Reprod Sci 2007; [Epub ahead of print].

[59] Aitken RJ, Wingate JK, De Iuliis GN, Koppers AJ, McLaughlin EA. Cis-unsaturated fatty acids stimulate reactive oxygen species generation and lipid peroxidation in human spermatozoa. *J. Clin. Endocrinol. Metab.* 2006; 91: 4154-63.

[60] Shukla S, Jha RK, Laloraya M, Kumar PG. Identification of non-mitochondrial NADPH oxidase and the spatio-temporal organization of its components in mouse spermatozoa. *Biochem. Biophys. Res. Commun.* 2005; 331: 476-83.

[61] Armstrong JS, Bivalacqua TJ, Chamulitrat W, Sikka S, Hellstrom WJ. A comparison of the NADPH oxidase in human sperm and white blood cells. *Int. J. Androl.* 2002; 25: 223-9.

[62] Albina JE, Caldwell MD, Henry WL, Mills CD. Regulation of macrophage functions by L-arginine. *J. Exp. Med.* 1989 169: 1021-9.

[63] Schuwerack PM, Lewis JW, Hoole D, Morley NJ. Ammonia-induced cellular and immunological changes in juvenile Cyprinus carpio infected with the blood fluke Sanguinicola inermis. *Parasitology* 2001; 122: 339-45.

[64] Minana MD, Corbalan R, Montoliu C, Teng CM, Felipo V. Chronic hyperammonemia in rats impairs activation of soluble guanylate cyclase in neurons and in lymphocytes: a putative peripheral marker for neurological alterations. *Biochem. Biophys. Res. Commun.* 1999; 257: 405-9.

[65] Guerrini VH. Excretion of ammonia by Lucilia cuprina larvae suppresses immunity in sheep. *Vet Immunol. Immunopathol.* 1997; 56: 311-7.

[66] Hardin-Pouzet H, Krakowski M, Bourbonniere L, Didier-Bazes M, Tran E, Owens T. Glutamate metabolism is down-regulated in astrocytes during experimental allergic encephalomyelitis. *Glia* 1997; 20: 79-85.

[67] Shapiro MS, DeCoursey TE. Selectivity and gating of the type L potassium channel in mouse lymphocytes. *J. Gen. Physiol.* 1991; 97: 1227-50.

[68] Cossack ZT, Prasad AS. Activities of purine catabolism related enzymes in zinc deficiency: relationship to T-lymphocyte dysfunction and hyperammonemia. *Int. J. Vitam. Nutr. Res.* 1991; 61: 51-6.

[69] Klucinski W, Targowski SP. Ammonia toxicity for mammalian and avian lymphocytes from blood. *Immunopharmacology* 1984; 8: 47-52.

[70] Targowski SP, Klucinski W, Babiker S, Nonnecke BJ. Effect of ammonia on in vivo and in vitro immune responses. *Infect. Immun.* 1984; 43: 289-93.

[71] Targowski SP, Klucinski W, Jaworek D. Effect of ammonia on viability and blastogenesis of bovine lymphocytes. *Vet Immunol. Immunopathol.* 1984; 5: 297-310.

[72] Mukhopadhyay NK, Saha AK, Smith W, Dowling JN, Hiserodt J, Glew RH. Inhibition of neutrophil and natural killer cell function by human seminal fluid acid phosphatase. *Clin. Chim. Acta* 1989; 182: 31-40.

[73] Kim SC, Kim HW. Effects of nitrogenous components of urine on sperm motility: an in vitro study. *Int. J. Androl.* 1998; 21: 29-33.

[74] Miller AL. Therapeutic considerations of L-glutamine: a review of the literature. *Altern Med. Rev.* 1999; 4: 239-48.

[75] Alevy YG, Mueller KR, Hutcheson P, Slavin RG. Immune response in experimentally induced uremia. V. Description of a model for chronic uremia in the rat that produces a long-term suppressive effect on T cell responses to mitogens. *J. Lab. Clin. Med.* 1983; 101: 717-25.

[76] Alevy YG, Mueller KR, Slavin RG. Immune response in experimentally induced uremia. IV. Characterization of suppressor peritoneal macrophages in the uremic rat. *J. Lab. Clin. Med.* 1982; 100: 735-44.

[77] Alevy YG, Hutcheson P, Mueller KR, Slavin RG. Suppressor alveolar macrophages in experimentally induced uremia. *J. Reticuloendothel. Soc.* 1983; 33: 11-20.

[78] Hoosein NM, Martin KJ, Abdul M, Logothetis CJ, Kaddurah-Daouk R. Antiproliferative effects of cyclocreatine on human prostatic carcinoma cells. *Anticancer Res.* 1995; 15: 1339-42.

[79] Martin KJ, Chen SF, Clark GM, Degen D, Wajima M, Von Hoff DD, Kaddurah-Daouk R. Evaluation of creatine analogues as a new class of anticancer agents using freshly explanted human tumor cells. *J. Natl. Cancer Inst.* 1994; 86: 608-13.

[80] Liu YX. Involvement of plasminogen activator and plasminogen activator inhibitor type 1 in spermatogenesis, sperm capacitation, and fertilization. *Semin Thromb Hemost* 2007; 33: 29-40.

[81] Ogetman Z, Dirlik M, Caglikulekci M, Canbaz H, Karabacak T, Yaylak F, Tamer L, Kanik A, Aydin S. The effect of aminoguanidine on blood and tissue lipid peroxidation in jaundiced rats with endotoxemia induced with LPS. *J. Invest. Surg.* 2006; 19: 19-30.

Ligands, Polymers and Amino Acids
Editor: Tsisana Shartava
ISBN: 978-1-61122-793-2

PREBIOSIS WITH FERMENTABLE CARBOHYDRATES IN PIG NUTRITION: A HOLISTIC APPROACH TO PRODUCTIVITY

Ajay Awati
The Riddet Institute, Massey University, Private bag 11222,
Palmerston North, New Zealand

ABSTRACT

Evolutionary processes over time have formed a symbiotic relationship between a well diverse and stable microbiota and the animal host. Berg [1] and Gaskins [2] have pointed out that bacterial cells outnumber animal cells by a factor of 10 and have a profound influence on nutritional, physiological, immunological, and protective processes in the host.

Essentially, gastrointestinal tracts of animals are sterile at birth. After the first introduction to microflora during birth, the GI tract gets colonized by microbes from mother and the environment, until a stable, diverse and complex microbiota exists. The successful development, stability, and functionality of this very complex ecosystem is dependent on the microbial community, host physiology and the diet (nutrient source for the ecosystem). Any alterations in these three pillars of the GIT microbial ecosystem, lead to changes in microbial activity with consequences on host physiology and performance.

This chapter is focused on the effects of prebiotics, essentially fermentable carbohydrates in the diet, on the gut microbial community, gut health, physiology and performance of the animal, and the environment in and around a pig farm. Author further discusses the candidature of fermentable carbohydrates as an alternative to in-feed antibiotics in pig production. Some recommendations for further research directions are also discussed.

INTRODUCTION

More than a decade ago, Gibson and Roberfroid [3], defined the term prebiotic as: 'a non digestible food ingredient that beneficially affects the host by selectively stimulating the growth and/or activity of one or a limited number of bacteria in the colon that can improve host health.' Emphasis in this definition on the colon may be too restrictive as many monogastric species, especially pigs [4] and sometimes humans support a considerable degree of bacterial fermentation in the upper digestive tract [5]. Additionally, in recent times focus has shifted towards the concept of the need for 'a stable and complex commensal bacterial community' as a pre-requisite for a healthy gut eco-system [6, 7]. Such considerations have led to further elaboration of the definition of a prebiotic by Awati [8] as 'a non digestible dietary ingredient that beneficially affects the host by stimulating the activity, in terms of fermentation end products, and stability of the diverse comensal microbiota in different parts of the gastro-intestinal tract, depending on the fermentability of the dietary ingredient itself.'

The sole purpose of a prebiotic from a physiological point of view is to provide a preferred substrate for the potentially beneficial microbiota.

Most of the compounds investigated for such prebiotic properties, have been dietary fibres. Soluble fibres are in general better energy substrates for gastrointestinal micro-organisms than are insoluble fibres [9].

The important role that prebiotics play in human nutrition as a functional food ingredient and their health benefits in general are well documented [10-19]. However, in pig nutrition, when viewed in the context of a production industry where efficiency of animal growth is a primary objective, prebiotics have a more diverse role to play. The effect of prebiotics on the gut microbial community and activity may be primary, but effects on overall performance, farm environment and farm economics are crucial as well.

EFFECTS AT GUT LEVEL

Microbial Community and Microbial Activity

The generalized theory behind the inclusion of fermentable carbohydrates in pig diets is to skew the fermentation process more towards carbohydrate fermentation as oppose to harmful protein fermentation. It is a well-documented fact that in the presence of the fermentable carbohydrate substrate, microbes prefer to ferment the carbohydrate source to derive energy and to use the nitrogen available for their own growth [20]. Protein fermentation is considered detrimental to the gut health due to the production of toxic substances such as ammonia, amines, skatol, indole etc. [21-23].

A study conducted on weaning piglets [24], in which two groups of animals were fed either of two semi-purified experimental diets, with or without the inclusion of fermentable carbohydrates, for 10 days post weaning, clearly demonstrated that the inclusion of fermentable carbohydrates in a diet significantly decreased ammonia concentration and branched chain proportion in fecal samples. The concentration of ammonia and branched chain proportion (BCP), which is described as the proportion of branched chain fatty acids of the total volatile fatty acids produced [25], are indicators of the extent of protein fermentation. The study concluded that fermentable carbohydrates skew the fermentation process away from the protein fermentation. Other studies have reported similar effects [26].

Different fermentable carbohydrates have different fermentation rates. Studies done using an *in vitro* gas production technique demonstrate differences in the fermentability of different carbohydrate substrates [27, 28]. It was hypothesized for the study by Awati *et al* [29], where weaning piglets were subjected to either of two dietary treatments with or without the inclusion of fermentable carbohydrates that fermentable carbohydrates with different fermentation rates would induce fermentation along the GI tract by providing the carbohydrate substrate for the microbiota in different parts of the GI tract. Fermentable carbohydrates included in the diet were lactulose, inulin, wheat starch (WST), and sugar beet pulp (SBP), with different fermentation rates in vitro (lactulose fastest to sugar beet pulp the slowest among four). After 10 days of feeding the experimental diets, the digesta samples were collected from different parts of gastro intestinal tract. It was observed that fermentation end product profiles in the small intestine as well as the large intestine were different between the two dietary groups. The group fed with diet inclusive of fermentable carbohydrates showed higher short chain fatty acid production and lower ammonia concentration and branched chain proportion in digesta samples. The study clearly demonstrated that by using fermentable carbohydrates with different fermentabilities, fermentation along the GIT can be skewed away from protein fermentation. In other words, the fermentability of the carbohydrate substrate determines the fate of substrate in terms of the site of fermentation in porcine GIT.

It is not only that fermentable carbohydrates change the microbial activity depending on their fermentability, but also exposure of microbiota for a certain period to a specific carbohydrate substrate can cause adaptation of the microbiota to that specific substrate. In an *in vitro* study [25], the effects of two different diets, for weaning piglets, on the microbial activities of fecal microbiota were compared between animals on the two different diets. The two diets were: CHO diet (containing added fermentable carbohydrates – including SBP and WST), and Control diet without any added fermentable carbohydrates. After nine days on the diet, fecal samples for inoculum were collected from selected animals, and tested for their activity in terms of gas production kinetics, and end-products such as volatile fatty acids (VFA), ammonia, and dry matter (DM) disappearance of two test substrates SBP and WST. The bacterial diversity was also analyzed in this study, before and after *in vitro* fermentation using denaturing gradient gel electrophoresis (DGGE) analysis of amplified 16S rRNA genes. There were differences both in the kinetics and in the end-products of the substrates. More interesting significant differences were detected between inocula, though mainly in terms of the fermentation kinetics of the two substrates. With the CHO fecal inoculum, SBP was fermented faster than WST, although it was expected that WST would ferment faster based on prior studies where fecal donor piglets were not exposed to any of the substrates (which was also observed in this study with the control group). The difference between the kinetics of SBP and WST fermentation by fecal microbiota from the CHO diet fed piglets, suggests better adaptation of the microbiota to SBP fermentation. This was probably because in the *in vivo* situation, given that a significant amount of starch is known to be fermentable by the small intestinal microbiota, WST was fermented in the proximal large intestine, while the fecal inoculum, which is representative of distal colonic microbiota, was adapted to the available SBP. Significantly higher diversity, as measured by DGGE fingerprint analysis, was detected in the microbial community enrichment by SBP as compared to WST at the end of fermentation. It was concluded that the microbial community composition and activity in the GIT may be changed in response to diet, and that this change was detected *in vitro*.

Konstantinov *et al.* [30] have shown the modulation of the bacterial community in the ileum and colon of weaning piglets after the inclusion of the fermentable carbohydrates namely, lactulose, inulin, wheat starch and sugar beet pulp. Comparative molecular microbiological analysis used in the study revealed that the principal microbiological difference between the ileal and colonic digesta was the occurrence of a higher number of DGGE bands in the colon compared to the ileum. The application of quantitative FISH (fluorescent in-situ hybridization) using a species specific probe revealed that the *Lactobacillus Amylovorus*-like population was the most prevalent in the ileal and colonic lumen samples of the piglets fed for 10 days with fermentable carbohydrates, compared to lumen samples of piglets fed a control diet with negligible amounts of fermentable substrates.

Various studies in pigs have reported the stimulation of beneficial commensal microbial species and increased diversity in gut microbiota [31-35].

The studies discussed here clearly point out that not all dietary fiber sources are equally fermentable, nor do they have similar effects on the microbial community and its activity. The fermentability (fermentation rate) of the carbohydrate substrate is the determining factor as to whether a particular dietary fiber will produce a prebiotic effect at the desired site in the gastro-intestinal tract.

Studies reported by Houdijk [36], also showed that the relatively fast fermenting fructo-oligosaccharides (FOS) affected microbial activity only in the small intestine of growing pigs. In a similar way, a study by Kamphues *et al.* [37] failed to get the desired prophylactic effect of lactulose on pigs experimentally infected with *Salmonella Derby*, probably because the lactulose was fermented so quickly that it did not reach to the desired site in the lower GIT.

Gut Development and Gut Health

Fermentable carbohydrates, especially by improving the microbial fermentation increase the production of short chain fatty acids (SCFA). SCFA, especially butyrate, produced by the microbiota during fermentation processes stimulate the development and growth of the large and the small intestine [38]. SCFA are an important source of energy for gut epithelium and help to maintain a balanced ecosystem. Fermentability of the carbohydrate substrate not only affects the amount but also the proportions of the different SCFA produced [39, 40]. Soluble fibers are more efficiently converted to butyric acid, which improves the epithelial cell proliferation in the small and large intestine (reviewed by [20]). SCFA are reported to stimulate the localised mucus secretion and thus add to the protection of gut epithelium [41].

Several studies have reported increased colon length and weight [42, 43], increased absolute and relative weight of caecum and colon [44], and increased villus height and crypt depth [42, 45] by the inclusion of fermentable fiber in the pig diet. Inclusion of FOS in neonatal pigs diet has shown preventive effect on colonic epithelial mucosa atrophy [31].

Apart from gut epithelial cell proliferation, fermentable carbohydrates, by influencing the growth of commensal beneficial microflora lead to colonization resistance [46] against pathogens. Colonization resistance, as reviewed by Williams *et al* [20], is a mechanism of species antagonism, involving indigenous microflora, by the microbial production of bacteriocins, VFA, extracellular enzymes, and competition for nutrients and attachment sites.

The literature is very contradictory about the use of dietary fibre in diet reduction in pathogenic infections. Recent studies on rats[47-49], have shown that dietary inclusion of FOS, inulin or lactulose, led to increased translocation of *Salmonella*, while there was a simultaneous decrease in the resistance to *Salmonella*. On the other hand, studies on poultry [50-52], have shown inhibition of Salmonella by the addition of complex carbohydrates. In pigs, certain studies concluded that limiting the fermentable substrates in GIT controls the proliferation of pathogenic bacteria [53-56] whereas other studies claim positive effects of fermentable carbohydrates against the pathogens. [57, 58]. In recent studies, Konstantinov *et al* [30], have shown, by DGGE analysis, the absence of bands of clostridia sp. in the ileal digesta samples from piglets, which had fermentable carbohydrates in their diets for 10 days post weaning. Reduced cecal density of clostridium is also reported in piglets fed a diet including lactulose [59].

Pie *et al* [60] reported that the fermentation end products profile is related to the inflammatory status of the gut. In this study, the concentrations of fermentation end products and inflammatory cytokines were correlated. The branched chain proportion and concentration of ammonia in digesta samples, which are dependent on protein fermentation, were positively correlated with pro-inflammatory cytokines IL-8, IL-12p40, and IL-18 mRNA. However, the group of piglets fed a diet containing fermentable carbohydrates showed increased levels of IL-6 mRNA. IL-6 is known to have anti inflammatory activities, such as the inhibition of macrophage TNF and IL-1 release. [61, 62]. Anti-inflammatory properties of prebiotics have also been reported by several other studies [63-65]

Interesting but conflicting studies have been reported regarding dietary fibres and the parasitic infestation of *Oesophagostomum dentatum* in the large intestine of growing pigs. Some of the earlier studies by Petkevicius *et al* [66-69], concluded that digestion-resistant fibers in diet provide a favourable conditions for the establishment of *O.dentatum*. In contrast, later studies, published by same author [44, 70], showed that the use of fermentable carbohydrates in the diet, actually had anti parasitic effects, and inhibited on *O. dentatum* establishment. Some other studies [71, 72] from the same group also demonstrated that the inclusion of the highly degradable fructose polymer inulin in the diet led to significant reductions in another important intestinal parasite *Trichuris suis* establishment, egg excretion, and female worm fecunditv. The authors even suggested, the possible use of fermentable carbohydrates as a treatment for parasitic infections [71]. They proposed that the effect was due to fermentation end products produced during fermentation of these fermentable carbohydrates, by the gut microbiota [73]. The gut microbial activity and SCFA produced during the fermentation of carbohydrate substrate has a protective effect on the gut by increased cell proliferation and increased mucin secretion [74]. The composition of the mucin is also known to be affected by fermentation process. These observations certainly support that the fermentability of the non digestible carbohydrate substrate being the determinant of the positive effects of dietary fibers on gut ecosystem.

EFFECTS ON ANIMAL

Animal Growth Performance

The literature is very confusing when it comes to the effects of fermentable carbohydrates on animal growth performance, which is conventionally measured in terms of feed intake, average daily weight gain and feed conversion ratio. The health benefits of fermentable carbohydrates are well accepted, but their negative effects on the growth performance have also been well documented [75-81]. Most studies showed, either reduced or no effect on growth performance by inclusion of fermentable carbohydrates in pig diets. However, it is interesting to note that these data are mainly from experimental farm studies, where the pathogenic microbial load is far less than in the conventional piggery and data in the respective studies are only recorded for a couple of weeks. Production traits measured for only couple of weeks in an animal study do not necessarily reflect the real effect of the dietary treatments (Personal communication Moughan PJ). A study by Houdijk [82], did not find any effect of adding non digestible oligosaccharides (NDO) in growing piglets' diet for six weeks. However, the authors observed an interesting phenomenon. There was an initial depression in feed intake and daily weight gain for first period of three weeks and then a compensatory increase in feed intake and daily weight gain. It was even speculated that the NDO fed group would have reached slaughter weight in fewer days than the control group, when the results were extrapolated in time. The authors theorised that, the ecology of the microbiota primes the immune system [83]. Any deviation from the primed situation triggers a non specific immune response. It is well accepted that the addition of a fermentable carbohydrate source in the diet causes changes in the microbiota and microbial activity. These changes subsequently cause a non specific immune response to commensal microbiota, leading to a depression in feed intake in the initial period of fermentable carbohydrate inclusion. However, soon after the adaptation to the dietary fermentable carbohydrate, feed intake, daily weight gain and other growth parameters show a compensatory increase. This is just one way of interpreting the result, and the lack of long term growth studies obscures the real effect of dietary fermentable carbohydrates on animal growth.

Nutrient Availability

Usually it is considered that increased fermentable carbohydrate content in the diet has a direct effect on nutrient and energy availability [84]. It is mainly due to the fact that higher amounts of fermentable carbohydrates in the diet increase water binding in the stomach and the viscosity of the chyme [85]. This reduces the feed intake of animal by causing earlier satiety by expansion of the stomach wall [86]. There is also the reduction in transit time of digesta in the upper gastrointestinal tract [87, 88], which reduces the nutrient availability.

However, the extensive review of the digestibility studies in pigs by Soffrant [89], is finally inconclusive about the effect of dietary fiber on ileal digestibility and endogenous nitrogen losses in the pig. Fermentable carbohydrates reduce the physical activity of the pig due to nutritional satiety, which contributes indirectly to positive energy balance [90].

In terms of energy contribution, fermentable carbohydrates are underestimated in monogastric nutrition. Volatile fatty acids produced during the microbial fermentation of carbohydrate substrates, contribute from 5 to 28 % (reviewed by Gaskins [4] and Bergman [91]) to the total body energy requirement of the pig. VFA, especially butyrate, provide 70% of the energy required by the colonic epithelium (reviewed by Williams *et al.* [20]). Furthermore, in the large intestine of a monogastric animal, fermentable carbohydrate is a limiting factor for growth and diversity of microbiota [92]. In the absence of carbohydrates, undigested protein is used as energy source by microbiota leading to production of ammonia. This ammonia, absorbed through the gut wall, must be excreted as urea with an additional cost of about 7% of total energy expenditure in animals [93]. However, in the presence of the carbohydrate substrate, the indigestible protein is used for microbial growth. With respect to significant contribution to the energy balance in pigs, Noblet and Le Goff [94] recommended to give at least two different energy values for the feeds containing fermentable carbohydrates: one for growing pigs and one for adult sows. In the latter case the positive contribution of dietary fiber is more pronounced [95].

Besides the energy contribution through VFA, microbiota also provide the host with nutrients including, vitamin K, vitamin B and some essential amino acids [96]. Short chain fatty acids also have positive effect on re-absorption of water from the colon maintaining the fluid balance of the host body [97]. There is increasing evidence in recent publications of improved mineral absorption and bone mineralization, with the addition of fermentable carbohydrates in the diets [98-105].

Apart from the above mentioned nutrient contributions from the microbial fermentation of fermentable carbohydrate substrates, some exciting information claiming microbial protein contributes to amino acid homeostasis of the monogastric host [106], is also published sporadically. However, this needs to be validated by more scientific evidence to quantify the extent.

Meat Production and Meat Quality

Boar taint, an off flavour in the meat, is one of the major issues in pork production. Skatole is responsible for boar taint in pigs [107]. Skatole is produced in the large intestine of the pig by microbial degradation of L-tryptophan originating from dietary and endogenous protein entering the large intestine [107]. This skatole is then absorbed in blood stream and increased levels of skatole into fat causes off flavour. The Inclusion of fermentable carbohydrates in the diet skew the fermentation process more into the carbohydrate direction and reduce the protein fermentation and so the skatole production. Several studies have reported the reduction in backfat skatole content by the inclusion of dietary fermentable carbohydrates [107, 108]. Intact males are known to give more boar taint than castrates [109, 110]. However, intact males have more lean meat and better feed conversion ratios than castrates. Consequently, the Inclusion of fermentable carbohydrates in the diets of intact male pigs [108] might improve the financial gain for farmers by better quality meat production. A study by Martelli *et al* [111] concluded that pressed beet pulp silage, either molassed or plain, can be profitably used in the diets for heavy pigs. The high energy supply of beet pulp, which is related to its non-starch fermentable polysaccharides content, can reduce feed cost by sparing cereals such as barley. A recent study with inulin rich dried chicory roots in diets of entire males and females reached a similar conclusion, that dried chicory may be the most suitable form for commercial use because it had no initial adverse effects on food intake, consistently reduced skatole without reducing performance, was easy to handle throughout the entire year and is relatively inexpensive [112]. Furthermore, fermentable carbohydrates provide a better option to deal with boar taint in organic pig farming, where castration of animals is not allowed.

Animal Behaviour and Well Being

The low feeding level in common practice for dry sows has been linked to the occurrence of stereotypical behaviours. These challenge the welfare of animal to a greater extent. A study by Brouns *et al.* [113] reported that feeding the fibrous diet once daily reduced the incidence of oral behaviours relative to control group, and the oral behaviours which still occurred did not have the appearance of stereotypies. Very few abnormal oral behaviours were observed when the fibrous diet was offered ad libitum. The time spent rooting was reduced when the diet was offered once daily, but was reduced even further when the diet was offered ad libitum. This suggests that the feeding motivation of gilts which received a fibrous diet was less than that of the control gilts. Several recent studies have reported these beneficial effects of fermentable carbohydrates on stereotypic behaviors in pregnant sows [114-120]. Guilleme *et al* [121] have even reported the improved piglets' growth rate during the 1st week of life and tended to increase their live weight at weaning after feeding sows with high fiber diet during gestation.

The nutritional and physical satiety caused by fermentable carbohydrates in the diet, reduce the stress on these animals and so the stereotypic behavior. Reduction in physical activity and stress due to the satiety has positive effect on the energy balance of the sows.

Some of these behavioral changes might be attributed to short chain fatty acids, especially butyric acid, produced during the fermentation in the gastrointestinal tract. Butyric acid can affect brain function [122].

Effects on Farm and Environment

Ammonia Emission

High intensification in animal production systems has witnessed an increasing public concern about environmental pollution [123]. Aarnink [124], noted that ammonia emission from surplus manure seems to create one of the most serious environmental problems. Globally, ammonia emission from livestock production is responsible for more than 50% of the total emission, of which 13% is from pig production [125]. Ammonia lost from manure not only decreases the fertiliser value, but also causes serious air pollution. Deposition of ammonia enhances the nitrogen enrichment of the soil, influencing the aquatic and terrestrial ecosystems. [126-128].

The effects of ammonia on indoor air quality in a pig farm, lead to serious health problems for pigs and have an effect on their energy metabolism, and so affect productivity. (reviewed by Canh [129]). It has been shown in several studies that increased fermentable carbohydrate levels in diets for pigs reduce the ammonia emission from slurry [130-134]. There are two probable main mechanisms behind this reducing effect. First, the increased fermentation, when feeding diets rich in fermentable carbohydrates, increases the excretion of SCFA in feces which decreases the pH-value in slurry [130-134].

The ammonia emission from slurry is directly related to the pH-value of the slurry, and lowering the pH decreases the emission. Secondly, nitrogen is repartitioned from urine to faecal microbial biomass [134, 135].

It has been suggested that there are two main factors contributing to the repartitioning of nitrogen from urine into feces when feeding diets rich in fermentable carbohydrates. First, more nitrogen and especially ammonia is incorporated into bacterial nitrogen and secondly more urea is diverted into the digestive tract due to an increased intestinal blood flow [136]. Nitrogen excreted in feces is mainly of bacterial origin (70–85%) [137, 138]. However, nitrogen excreted in urine is mainly in the form of urea, which is more susceptible to decomposition than bacterial nitrogen excreted in feces [139]. Consequently, the repartitioning of nitrogen results in a slurry with a less decomposable nitrogen content.

This negative effect on ammonia emission makes the fermentable carbohydrates a worthwhile dietary component in pig nutrition, especially in countries where there is strict legislation to reduce the emission of ammonia to acceptable levels.

Farm Odor

Odor coming from pig slurries caused an increased public opposition to intensive pig farms. A recent report (http://bankwatch.org/documents/smithfield_report_september2006.pdf) has summarized the potential effects of pig farm odors on the general wellbeing of populations including farm workers, living near pig farms. The odor affects both physiological and psychological health. The symptoms include nausea, headaches, shallow breathing, coughing, sleep disturbance, and loss of appetite. In addition to odour malfeasance, neighbours appear to be experiencing elevated rates of health symptoms related to the upper respiratory tract as well as immune system damage.

Numerous odorous compounds have been identified in the pig farm wastes. However, in an intensive pig unit the primary malodorous compounds were associated with branched chain fatty acids, phenol, p-cresol, indole, and skatole [140], which are primarily end products of protein fermentation. As discussed earlier in this chapter, the volatile compounds like ammonia and the decomposition of urea nitrogen from urine, also contribute to the odour. Kaufmann [141] proposed that lack of specific fermentable non-starch polysaccharide levels or excessive protein in the large intestine and an increased pH level in cecal and colonic chyme will produce increased levels of the odorous compounds. Bunce *et al.* [142] showed that with inclusion of fermentable carbohydrates in diet the excretion of odorous compounds is reduced. Sutton *et al.* [143] discussed the possibilities of reducing of odorous compounds in swine manure through diet modification. Increasing the fermentable carbohydrates and reducing the indigestible protein entering large intestine would be the best bet. Fermentable carbohydrates certainly offer an avenue in the countries with increasing legal burden and penalties for odor from pig farms.

Fermentable Carbohydrates as Potential Alternatives to Anti-Microbial Growth Promoters (AMGP)

In last few decades, feeding pigs with inclusion of sub-therapeutic levels of antibiotics in diets has been done to achieve higher weight gain and feed efficiency. Although the exact mechanism behind this has not been well understood, effects were more pronounced on the farms with poor hygiene status. There are different opinions about the mechanisms;

a) Inhibition of pathogenic bacteria [144, 145]
b) Prevention of irritation of the intestinal lining and enhanced uptake of nutrients from the intestine by thinning of the mucosal layer [145, 146]

c) Increase levels of insulin like growth factor [147, 148] shown by some antibiotic agents.

Despite these good reasons, a ban on in-feed antibiotics in pig production systems is inevitable, given the increasing worldwide concern about antibiotic resistance in human and veterinary medicine [149-152].

The ban on in-feed antibiotics decreased the antibiotic resistance in pig production systems of Denmark and Sweden [153-155]. However, there was a sudden increase in use of therapeutic antibiotics in Denmark, while Swedish farmers managed to lower the use of antibiotics, by improving the husbandry practices and hygiene on the farm. However, this in turn would have increased the production cost of the pig meat.

This implies that any alternative to AMGP should fill the gap by reducing the incidence of pathogenic infections and keep the economics of pig production sustainable.

Traditionally, pig production performance has always been measured in terms of body weight gain, feed intake and feed conversion. An alternative approach of looking at the overall production cost of pig meat might give more insight the criteria for selection of an alternative to AMGP.

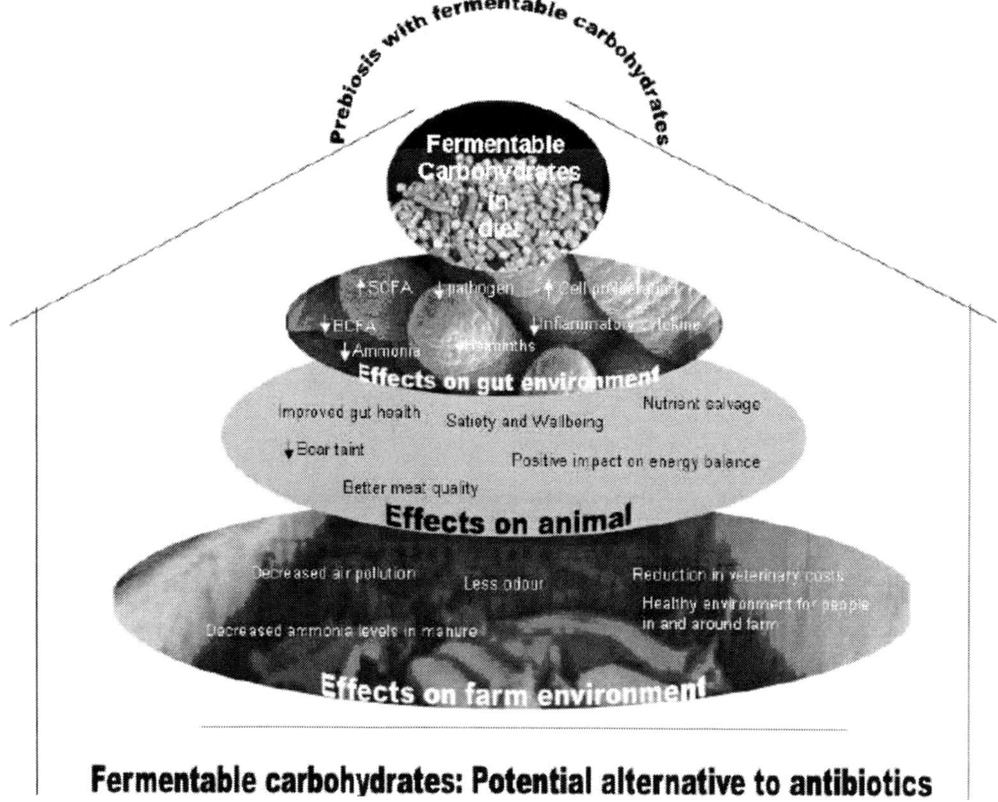

Figure1. Beneficial effects of prebiosis with fermentable carbohydrates at gut, animal and farm level, supporting the fermentable carbohydrates as potential alternative to in-feed antibiotics.

Fermentable carbohydrates have this potential based on following issues discussed in this chapter (see figure1):

- Reduction in veterinary cost involved (discussed by Houdijk [36]) due to improved gut health with lower pathogenic and parasitic load.
- Reduction in environmental cost by reduction in urinary N excretion (which is already a serious issue in western countries)
- Reduction in public defiance against pig farms by reduction in excretion of odor metabolites, which is also important for health of pigs and farm workers.
- Reduction in boar taint in meat and improvement in meat quality.
- Nutrient salvage and availability
- Reduction in stereotypic behavior resulting in more welfare of the animals.

Conclusion

The issues discussed in this chapter certainly indicate the overall importance of fermentable carbohydrates in pig nutrition. However, It is necessary to conduct the long term studies which will study the socio-economic aspects of inclusion of various fermentable carbohydrates in pig diets, at each step of pig production system......from farm to the fork!!

References

[1] Berg, R. D. (1996). The indigenous gastrointestinal microflora. *Trends in Microbiology, 4(11), 430-435.*

[2] Gaskins, H. R. (1997). Immunological aspects of host/microbiota interactions at the intestinal epithelium. In R. I. Mackie, B. A. White, and R. E. Isaacson, (Eds.), *Gastrointestinal Microbiology* (142-186). New York: Chapman and Hall.

[3] Gibson, G. R. and Roberfroid M. B. (1995). Dietary modulation of the human colonic microbiota: Introducing the concept of prebiotics. *Journal of Nutrition, 125, 1401-1412.*

[4] Gaskins, H. R. (2001). Intestinal bacteria and their influence on swine growth. In A. J. Lewis and L. L. Southern (Eds.), *Swine Nutrition (585-608).* Florida: CRC Press LLC.

[5] Wilson, K. H. (1997). Biota of human gastrointestinal tract. In R. I. Mackie, B. A. White, and R. E. Isaacson, (Eds.), *Gastrointestinal Microbiology* (39-58). New York: Chapman and Hall.

[6] Verstegen, M. W. A. and Williams, B. A. (2002). Alternatives to the use of antibiotics as growth promoters for monogastric animals. *Animal Biotechnology, 113-127.*

[7] Konstantinov, S. R., De Vos, W. M., Akkermans, A. D. L., Smidt H., Favier C. F., Zhu W. Y., Williams B. A., Kluss J., and Souffrant, W-B (2004). Microbial diversity studies of the porcine gastrointestinal ecosystem during weaning transition. *Animal Research, 317-324.*

[8] Awati, A. (2005). Prebiotics in piglet nutrition? fermentation kinetics along the GI tract. PhD thesis Animal nutrition group, Wageningen University, Wageningen, The Netherlands.

[9] Fahey, J. G. C., Flickinger, E. A., Grieshop, C. M., and Swanson, K. S. (2004). The role of dietary fibre in companion animal nutrition. In J. W. Kamp van der, N. G. Asp, J. Miller Jones, and G. Schaafsma (Eds.), *Dietary Fibre: bio-active carbohydrates for*

food and feed (295-328). Wageningen, The Netherlands: Wageningen Academic Publisher.

[10] Andrieux, C. (2001). Prebiotics and health. In *Post-Antibiotics era of animal nutrition: The 3rd Techno World Meeting*. Korea: CTC-BIO.

[11] Bengmark, S. (2002). Gut microbial ecology in critical illness: Is there a role for prebiotics, probiotics, and synbiotics?. *Current Opinion in Critical Care, 8, 145-151.*

[12] Berg, R. D. (1998). Robiotics, prebiotics or 'conbiotics'? *Trends in Microbiology, 6(3), 89-92.*

[13] Blaut, M. (2002). Relationship of prebiotics and food to intestinal microflora. *European Journal Nutrition, 41(Suppl 1), I11-I16.*

[14] Chow, J. (2002). Probiotics and prebiotics: A brief overview. *Journal of Renal Nutrition, 12(2), 76-86.*

[15] Collins, M. D. and Gibson G. R. (1999). Probiotics, prebiotics and synbiotics: approaches for modulating the microbial ecology of the gut. *American journal of clinical nutrition, 69 (suppl), 1052S-1057S.*

[16] Fooks, L. J., Fuller R., and Gibson G. R. (1999). Prebiotics, probiotics and human gut microbiology. *International dairy journal, 9, 53-61.*

[17] Macfarlane, G. T. and Cummings J. H. (1999). Probiotics and prebiotics: can regulating the activities of intestinal bacteria benefit health? *British Medical Journal, 318, 999-1003.*

[18] Rastall, R. A. and V. Maitin (2002). Prebiotics and synbiotics: towards the next generation. *Current Opinion in Biotechnology, 13(5), 490-496.*

[19] Roberfroid, M. (2002). Functional food concept and its application to prebiotics. *Digestive and Liver Disease, 34 (Suppl 2), S105-110.*

[20] Williams, B. A., Verstegen, M. W. A. and Tamminga, S. (2001) Fermentation in the large intestine of single-stomached animals and its relationship to animal health. *Nutrition Research Reviews, 14, 207-227.*

[21] Yokoyama, M. T., Tabori, C., Miller E. R., and Hogberg, M. G (1982). The effects of antibiotics in the weanling pig diet on growth and the excretion of volatile phenolic and aromatic bacterial metabolites. *American Journal of Clinical Nutrition, 35(6), 1417-1424.*

[22] Russell, J. B., Sniffen, C. J. and Van Soest, P. J. (1983). Effect of carbohydrate limitation on degradation and utilization of casein by mixed rumen bacteria. *Journal of Dairy Science, 66, 763-775.*

[23] Macfarlane, G. T., Gibson, G. R., Beatty, E., and Cummings, J. H. (1992). Estimation of short-chain fatty acid production from protein by human intestinal bacteria based on branched-chain fatty acid measurements. *FEMS Microbiology ecology, 101(2), 81-88.*

[24] Awati, A., Williams, B. A., Bosch, M. W., Gerrits, W. J. J., and Verstegen, M. W. A. (2006). Effects of inclusion of fermentable carbohydrates in the diet on fermentation end product profile in feces of weanling piglets. *Journal of Animal Science, 84, 2133-2140.*

[25] Awati, A., Konstantinov, S. R., Williams, B. A., Akkermans, A. D. L., Bosch, M. W., Smidt, H, and Verstegen, M. W. A. (2005). Effect of substrate adaptation on the microbial fermentation and microbial composition of faecal microbiota of weaning piglets studied in vitro. *Journal of the Science of Food and Agriculture, 85(10), 1765-1772.*

[26] Bikker, P., Dirkzwager, A., Fledderus, J., Trevisi, P., Luron, I., Lalles, J. P., and Awati, A. (2006). The effect of dietary protein level and fermentable carbohydrates level in newly weaned pigs on performance and intestinal characteristics. *Journal of Animal Science, 84, 3337-3345.*

[27] Williams, B. A., Bosch, M. W., Awati, A., Konstantinov, S. R., Smidt, H., Akkermans, A. D. L., Verstegen, M. W. A., and Tamminga S. (2005). In vitro assessment of gastrointestinal tract (GIT) fermentation in pigs: Fermentable substrates and microbial activity. *Animal Research, 54(3), 191-201.*

[28] Williams, B. A., Bosch, M. W., Boer, H., Verstegen, M. W. A., and Tamminga, S. (2005). An in vitro batch culture method to assess potential fermentability of feed ingredients for monogastric diets. *Animal Feed Science and Technology, 123-124(Part 1), 445-462.*

[29] Awati, A., Williams, B. A., Bosch, M. W., and Verstegen, M. W. A. (2006). Dietary carbohydrates with different rates of fermentation affect fermentation end- product profiles in different sites of gastrointestinal tract of weaning piglet. *Animal Science, 82, 837-843.*

[30] Konstantinov, S. R., Awati, A., Smidt, H., Williams, B. A., Akkermans, A. D. L., and De Vos, W. M (2004). Specific response of a novel and abundant Lactobacillus amylovorus-like phylotype to dietary prebiotics in the guts of weaning piglets. *Applied and Environmental Microbiology, 70(7), 3821-3830.*

[31] Howard, M. D., Gordon, D. T., Pace, L. W., Garleb, K. A., and Kerley, M. S. (1995). Effects of Dietary Supplementation with Fructooligosaccharides on Colonic Microbiota Populations and Epithelial-Cell Proliferation in Neonatal Pigs. *Journal of Pediatric Gastroenterology and Nutrition, 21(3), 297-303.*

[32] Smiricky-Tjardes, M. R., Grieshop, C. M., Flickinger, E. A., Bauer, L. L., and Fahey, G. C. (2003). Dietary galactooligosaccharides affect ileal and total-tract nutrient digestibility, ileal and fecal bacterial concentrations, and ileal fermentative characteristics of growing pigs. *Journal of Animal Science, 81(10), 2535-2545.*

[33] Konstantinov, S. R., Zhu, W. Y., Williams, B. A., Tamminga, S., de Vos, W. M., and Akkermans, A. D. L. (2003). Effect of fermentable carbohydrates on piglet faecal bacterial communities as revealed by denaturing gradient gel electrophoresis analysis of 16S ribosomal DNA. *FEMS microbiology ecology, 43:225-235.*

[34] Bikker, P., Dirkzwager, A., Fledderus, J., Trevisi, P., Huerou-Luron, l., Lalles, J. P., and Awati, A. (2007). Dietary protein and fermentable carbohydrates contents influence growth performance and intestinal characteristics in newly weaned pigs. *Livestock Science, 108(1-3), 194-197.*

[35] Roca-Canudas, M., Anguita, M., Nofrarias, M,. Majo, N., de Rozas, A. M. P., Martin-Orue, S. M., Perez, J. F., Pujols, J., Segales, J., and Badiola, I. (2007). Effects of different types of dietary non-digestible carbohydrates on the physico-chemical properties and microbiota of proximal colon digesta of growing pigs. *Livestock Science, 109(1-3), 85-88.*

[36] Houdijk, J.G.M. (1998). Effect of non-digestible oligosaccharides in young pig diets. PhD thesis Animal nutrition group, Wageningen University, Wageningen, The Netherlands.

[37] Kamphues, J., Tabeling, R., Stuke, O., Bollmann, S., and Amtsberg, G. (2007). Investigations on potential dietetic effects of lactulose in pigs. *Livestock Science, 109(1-3), 93-95.*

[38] Sakata, T. (1987). Stimulatory Effect of Short-Chain Fatty-Acids on Epithelial-Cell Proliferation in the Rat Intestine - a Possible Explanation for Trophic Effects of Fermentable Fiber, Gut Microbes and Luminal Trophic Factors. *British Journal of Nutrition, 58(1), 95-103.*

[39] Macfarlane, G.T. and Macfarlane S. (1993). Factors affecting fermentation reactions in the large bowel. *Proceedings of the Nutrition Society, 52(2), 367-373.*

[40] Macfarlane, S. and Macfarlane, G.T. (2003). Regulation of short chain fatty acid production. *Proceedings of the Nutrition Society, 62, 67-72.*

[41] Sakata, T. and Setoyama, H. (1995). Local Stimulatory Effect of Short-Chain Fatty-Acids on the Mucus Release from the Hindgut Mucosa of Rats (Rattus-Norvegicus). *Comparative Biochemistry and Physiology, 111(3), 429-432.*

[42] Hedemann, M. S. and Bach Knudsen, K. E. (2007). Resistant starch for weaning pigs-Effect on concentration of short chain fatty acids in digesta and intestinal morphology. *Livestock Science, 108(1-3), 175-177.*

[43] Martinez-Puig, D., Perez, J. F., Castillo, M., Andaluz, A., Anguita, M., Morales, J., and Gasa, J. (2003). Consumption of raw potato starch increases colon length and fecal excretion of purine bases in growing pigs. *Journal of Nutrition, 133, 134-139.*

[44] Petkevicius, S., Knudsen, K. E. B., and Murrell, K. D. (2003). Effects of Oesophagostomum dentatum and dietary carbohydrates on morphology of the large intestine of pigs. *Veterinary Parasitology, 116(2), 125-138.*

[45] Marinho, M. C., Pinho, M. A., Mascarenhas, R. D., Silva, F. C., Lordelo, M. M., Cunha, L. F., and Freire, J. P. B. (2007). Effect of prebiotic or probiotic supplementation and ileo rectal anastomosis on intestinal morphology of weaned piglets. *Livestock Science, 108(1-3), 240-243.*

[46] Vanderwaaij, D. (1987). Colonization Resistance of the Digestive-Tract - Mechanism and Clinical Consequences. *Nahrung-Food, 31(5-6), 507-517.*

[47] Bovee-Oudenhoven, I. M. J., Bovee-Oudenhoven, I. M. J., Ten Bruggencate, S. J. M., Lettink-Wissink, M. L. G., and Van Der Meer, R. (2003). Dietary fructo-oligosaccharides and lactulose inhibit intestinal colonisation but stimulate translocation of salmonella in rats. *Gut ,1572-1578.*

[48] Bovee-Oudenhoven, I. M. J., Ten Bruggencate, S. J. M., Bovee-Oudenhoven, I. M. J., Lettink-Wissink, M. L. G., Katan, M. B., and Van Der Meer, R. (2004) Dietary fructo-oligosaccharicles and inulin decrease resistance of rats to salmonella: Protective role of calcium. *Gut, 530-535.*

[49] Ten Bruggencate, S. J. M., Bovee-Oudenhoven, I. M. J., Lettink-Wissink, M. L. G., and Van der Meer, R. (2003). Dietary fructo-oligosaccharides dose-dependently increase translocation of salmonella in rats. *Journal of Nutrition, 2313-2318.*

[50] Fukata, T., Sasai, K., Miyamoto, T., and Baba, E. (1999). Inhibitory effects of competitive exclusion and fructooligosaccharide, singly and in combination, on Salmonella colonization of chicks. *Journal of Food Protection, 229-233.*

[51] Oyarzabal, O. A. and Conner, D. E. (1995). In vitro fructooligosaccharide utilization and inhibition of Salmonella spp. by selected bacteria. *Poultry science, 1418-1425.*

[52] Spencer, J. L., Chambers, J. R., and Modler, H. W. (1998). Competitive exclusion of Salmonella typhimurium in broilers fed with vermicompost and complex carbohydrates. *Avian Pathology, 244-249.*

[53] Montagne, L., Pluske, J. R. and Hampson D. J. (2003). A review of interactions between dietary fibre and the intestinal mucosa, and their consequences on digestive health in young non-ruminant animals. *Animal Feed Science and Technology, 95-117.*

[54] Pluske, J. R., Siba, P. M., Pethick, D. W., Durmic, Z., Mullan, B. P., and Hampson, D. J. (1996). The incidence of swine dysentery in pigs can be reduced by feeding diets that limit the amount of fermentable substrate entering the large intestine. *Journal of Nutrition 2920-2933.*

[55] Montagne, L., Lallès, J. P., Cavaney, F. S., Hampson, D. J., and Pluske, J. R. (2004). Effect of diet composition on postweaning colibacillosis in piglets. *Journal of Animal Science, 82(8), 2364-2374.*

[56] Hampson, D. and Pluske, J. (2004). Role of diet in managing enteric disease in pigs. *In Practice*, 438-443.

[57] May, T., Mackie, R. I., Fahey, J. G. C., Cremin, J. C., and Garleb, K. A. (1994). Effect of fiber source on short-chain fatty acid production and on the growth and toxin production by clostridium difficile. *Scandinavian Journal of Gastroenterology, 916-922.*

[58] Prohaszka, L. (1986). Antibacterial mechanism of volatile fatty acids in the intestinal tract of pigs against Escherichia coli. Zentralblatt fur Veterinarmedizin. Reihe B. *Journal of veterinary medicine. Series B, 166-173.*

[59] Kien, C. L., Blauwiekel, R., Williams, C. H., Bunn, J. Y., and Buddington, R. K. (2007). Lactulose feeding lowers cecal densities of clostridia in piglets. *Journal of Parenteral and Enteral Nutrition, 31(3),194-198.*

[60] Pie, S., Awati, A., Vida, S., Falluel, I., Williams, B. A., and Oswald, I. P. (2007). Effects of added fermentable carbohydrates in the diet on intestinal proinflammatory cytokine-specific mRNA content in weaning piglets. *Journal of Animal Science, 85(3), 673-683.*

[61] Akira, S., Hirano, T., Taga, T., and Kishimoto, T. (1990). Biology of Multifunctional Cytokines - Il-6 and Related Molecules (Il-1 and Tnf). *Faseb Journal, 4(11), 2860-2867.*

[62] Van Snick, J. (1990). Interleukin-6: An Overview. *Annual Review of Immunology, 8, 253-278.*

[63] Kanauchi, O., Nakamura, T., Agata, K., Mitsuyama, K., and Iwanaga, T. (1998). Effects of germinated barley foodstuff on dextran sulfate sodium-induced colitis in rats. *Journal of Gastroenterology, 33(2),179-188.*

[64] Videla, S., Vilaseca, J., Antolin, M., Garcia-Lafuente, A., Guarner, F., Crespo, E., Casalots, J., Salas, A., and Malagelada, J. R. (2001). Dietary inulin improves distal colitis induced by dextran sodium sulfate in the rat. *American Journal of Gastroenterology, 96(5), 1486-1493.*

[65] Cherbut, C., Michel, C., and Lecannu, G. (2003). The prebiotic characteristics of fructooligosaccharides are necessary for reduction of TNBS-induced colitis in rats. *Journal of Nutrition, 133(1), 21-27.*

[66] Petkevicius, S., Knudsen, K. E. B., Nansen, P., and Murrell, K. D. (2001). The effect of dietary carbohydrates with different digestibility on the populations of Oesophagostomum dentatum in the intestinal tract of pigs. *Parasitology, 123, 315-324.*

[67] Petkevicius, S., Nansen, P., Knudsen, K. E. B., and Skjoth, F. (1999). The effect of increasing levels of insoluble dietary fibre on the establishment and persistence of Oesophagostomum dentatum in pigs. *Parasite-Journal De La Societe Francaise De Parasitologie, 6(1), 17-26.*

[68] Petkevicius, S., Knudsen, K. E. B., Nansen, P., Roepstorff, A., Skjoth, F., and Jensen, K. (1997). The impact of diets varying in carbohydrates resistant to endogenous enzymes and lignin on populations of Ascaris suum and Oesophagostomum dentatum in pigs. *Parasitology, 114, 555-568.*

[69] Petkevicius, S., Bjorn, H., Roepstorff, A., Nansen, P., Knudsen, K. E. B., Barnes, E. H., and Jensen, K. (1995). The effect of two types of diet on populations of Ascaris suum and Oesophagostomum dentatum in experimentally infected pigs (vol 111, pg 395, 1995). *Parasitology, 111, 656-656.*

[70] Petkevicius, S., Knudsen, K. E. B., Murrell, K. D., and Wachmann, H (2003). The effect of inulin and sugar beet fibre on Oesophagostomum dentatum infection in pigs. *Parasitology, 127, 61-68.*

[71] Petkevicius, S., Thomsen, L. E., Knudsen, K. E. B., Murrell, K. D., Roepstorff, A., and Boes, J. (2007). The effect of inulin on new and on patent infections of Trichuris suis in growing pigs. *Parasitology, 134, 121-127.*

[72] Thomsen, L. E., Petkevicius, S., Knudsen, K. E. B., and Roepstorff, A. (2005). The influence of dietary carbohydrates on experimental infection with Trichuris suis in pigs. *Parasitology, 131, 857-865.*

[73] Petkevicius, S., Murrell, K. D., Knudsen, K. E. B., Jorgensen, H., Roepstorff, A., Laue, A., and Wachmann, H. (2004). Effects of short-chain fatty acids and lactic acids on survival of Oesophagostomum dentatum in pigs. *Veterinary Parasitology, 122(4), 293-301.*

[74] Libao-Mercado, A. J., Zhu, C. L., Fuller, M. F., Rademacher, M., Seve, B., and de Lange, C. F. M. (2007). Effect of feeding fermentable fiber on synthesis of total and mucosal protein in the intestine of the growing pig. *Livestock Science, 109(1-3),125-128.*

[75] Durmic, Z., Pethick, D. W., Pluske, J. R., and Hampson, D. J. (1998). Changes in bacterial populations in the colon of pigs fed different sources of dietary fibre, and the development of swine dysentery after experimental infection. *Journal of Applied Microbiology, 574-582.*

[76] Hopwood, D. E., Pethick, D. W., Pluske, J. R., and Hampson, D. J. (2004). Addition of pearl barley to a rice-based diet for newly weaned piglets increases the viscosity of the intestinal contents, reduces starch digestibility and exacerbates post-weaning colibacillosis. *British journal of nutrition, 419-427.*

[77] McDonald, D. E., Pethick, D. W., Pluske, J. R., and Hampson, D. J. (1999). Adverse effects of soluble non-starch polysaccharide (guar gum) on piglet growth and experimental colibacillosis immediately after weaning. *Research in Veterinary Science, 245-250.*

[78] Pluske, J. R., Black, B., Pethick, D. W., Mullan, B. P., and Hampson, D. J. (2003). Effects of different sources and levels of dietary fibre in diets on performance, digesta

characteristics and antibiotic treatment of pigs after weaning. *Animal Feed Science and Technology,129-142.*

[79] Pluske, J. R., Durmic, Z., Pethick, D. W., Mullan, B. P., and Hampson, D. J. (1998). Confirmation of the role of rapidly fermentable carbohydrates in the expression of swine dysentery in pigs after experimental infection. *Journal of Nutrition, 1737-1744.*

[80] Pluske, J. R., Pethick, D. W., Hopwood, D. E., and Hampson, D. J. (2002). Nutritional influences on some major enteric bacterial diseases of pigs. *Nutrition Research Reviews, 333-371.*

[81] Pluske, J.R., Pethick, D. W., and Mullan, B. P. (1998). *Differential effects of feeding fermentable carbohydrate to growing pigs on performance, gut size and slaughter characteristics. Animal Science, 147-156.*

[82] Houdijk, J. G. M., Bosch, M. W., Verstegen, M. W. A., and Berenpas, H. J. (1998). Effects of dietary oligosaccharides on the growth performance and faecal characteristics of young growing pigs. *Animal Feed Science and Technology, 71(1-2), 35-48.*

[83] Tannock, G.W. (1990). The Microecology of Lactobacilli Inhabiting the Gastrointestinal-Tract. *Advances in Microbial Ecology, 11, 147-171.*

[84] Wenk, C. (2001). The role of dietary fibre in the digestive physiology of the pig. *Animal Feed Science and Technology, 90,21-33.*

[85] Johansen, H. N., Knudsen, K. E. B., Sandstrom, B., and Skjoth, F. (1996). Effects of varying content of soluble dietary fibre from wheat flour and oat milling fractions on gastric emptying in pigs. *British Journal of Nutrition, 75(3), 339-351.*

[86] Langhans, W. (1999). Appetite regulation. In G.E. Lobley, A. White, and J.C. MacRae (Eds), *Protein Metabolism and Nutrition, EAAP Publication number 96*, Wageningen, The Netherlands: Wageningen press.

[87] Jorgensen, H., Zhao Xin, Q., and Eggum Bjorn, O. (1996). The influence of dietary fibre and environmental temperature on the development of the gastrointestinal tract, digestibility, degree of fermentation in the hind-gut and energy metabolism in pigs. *British Journal of Nutrition, 75(3), 365-378.*

[88] van Leeuwen, P. and Jansman, A. J. M. (2007). Effects of dietary water holding capacity and level of fermentable organic matter on digesta passage in various parts of the digestive tract in growing pigs. *Livestock Science, 109(1-3), 77-80.*

[89] Souffrant, W. B. (2001). Effect of dietary fibre on ileal digestibility and endogenous nitrogen losses in the pig. *Animal Feed Science and Technology, 90, 93-102.*

[90] Rijnen, M. M. J. A., Heetkamp, M. J. W., Schrama, J. W., Verstegen, M. W. A., Haaksma, J. (2001). Effects of dietary fermentable carbohydrates on energy metabolism in group-housed sows. *Journal of Animal Science, 148-154.*

[91] Bergman, E.N. (1990) Energy contributions of volatile fatty acids from the gastrointestinal tract in various species. *Physiological Reviews, 70(2), 567-590.*

[92] Rabiu, B. A. and Gibson, G. R. (2002). Carbohydrates: a limit on bacterial diversity within the colon. *Biol Rev Camb Philos Soc, 77(3), 443-453.*

[93] Eisemann, J. H. and Nienaber, J. A. (1990). Tissue and whole body oxygen uptake in fed and fasted steers. *British Journal of Nutrition, 64, 399-411.*

[94] Noblet, J. and Goff, G. L. (2001). Effect of dietary fibre on the energy value of feeds for pigs. *Animal Feed Science and Technology, 90, 35-52.*

[95] Muller, H. L. and Kirchgessner, M. (1985). Energy-Utilization of Pectin by Adult Sows. *Zeitschrift Fur Tierphysiologie Tierernahrung Und Futtermittelkunde-Journal of Animal Physiology and Animal Nutrition, 54(1), 14-20.*

[96] Savage, D.C. (1986). Gastrointestinal Microflora in Mammalian Nutrition. *Annual Review of Nutrition, 6, 155-178.*

[97] Ruppin, H., Bar-Meir, S., and Soergel, K. H. (1980). Absorption of short-chain fatty acids by the colon. *Gastroenterology, 1500-1507.*

[98] Perez-Conesa, D., Lopez, G., and Ros, G. (2007). Effects of probiotic, prebiotic and synbiotic follow-up infant formulas on large intestine morphology and bone mineralisation in rats. *Journal of the Science of Food and Agriculture, 87(6), 1059-1068.*

[99] Scholz-Ahrens, K. E., Ade, P., Marten, B., Weber, P., Timm, W., Asil, Y., Gluer, C. C., and Schrezenmeir, J. (2007). Prebiotics, probiotics, and synbiotics affect mineral absorption, bone mineral content, and bone structure. *Journal of Nutrition, 137(3), 838S-846S.*

[100] Perez-Conesa, D., Lopez, G., and Ros, G. (2007). Effect of probiotic, prebiotic and synbiotic follow-up infant formulas on iron bioavailability in rats. *Food Science and Technology International, 13(1), 69-77.*

[101] Perez-Conesa, D., Lopez, G., Abellan, P., and Ros, G. (2006). Bioavailability of calcium, magnesium and phosphorus in rats fed probiotic, prebiotic and synbiotic powder follow-up infant formulas and their effect on physiological and nutritional parameters. *Journal of the Science of Food and Agriculture, 86(14), 2327-2336.*

[102] Bosscher, D., Van Loo, J., and Franck, A. (2006). Inulin and oligofructose as functional ingredients to improve bone mineralization. *International Dairy Journal, 16(9), 1092-1097.*

[103] Franck, A. (2005). Prebiotics stimulate calcium absorption: a review. *Food Australia, 57(12), 530-532.*

[104] Abrams, S. A., Griffin, I. J., Hawthorne, K. M., Liang, L., Gunn, S. K., Darlington, G., and Ellis, K. J. (2005). A combination of prebiotic short- and long-chain inulin-type fructans enhances calcium absorption and bone mineralization in young adolescents. *American Journal of Clinical Nutrition, 82(2), 471-476.*

[105] Yeung, C. K., Glahn, R. P., Welch, R. M., and Miller, D. D. (2005). Prebiotics and iron Bioavailability - Is there a connection? *Journal of Food Science, 70(5), R88-R92.*

[106] Metges, C. C. (2000). Contribution of microbial amino acids to amino acid homeostasis of the host. *Journal of Nutrition, 130, 1857S-1864S.*

[107] Jensen, M. T., Cox, R. P. and Jensen, B. B. (1995). Microbial-Production of Skatole in the Hind Gut of Pigs Given Different Diets and Its Relation to Skatole Deposition in Backfat. *Animal Science, 61, 293-304.*

[108] Knarreborg, A., Beck, J., Jensen, M. T., Laue, A., Agergaard, N., and Jensen, B. B. (2002). Effect of non-starch polysaccharides on production and absorption of indolic compounds in entire male pigs. *Animal Science, 74, 445-453.*

[109] Babol, J., Squires, E. J., and Gullett, E. A. (1995). Investigation of factors responsible for the development of boar taint. *Food Research International, 28(6), 573-581.*

[110] Babol, J. and Squires, E. J. (1995). Quality of Meat from Entire Male Pigs. *Food Research International, 28(3), 201-212.*

[111] Martelli, G., Parisini, P., Sardi, L., Scipioni, R., Vignola, G., Panciroli, A., and Mordenti, A. (1999). The effects of sugar beet pulp silage in the diets for heavy pigs. *Annales De Zootechnie, 48(3), 173-188.*

[112] Hansen, L. L., Mejer, H., Thamsborg, S. M., Byrne, D. V., Roepstorff, A., Karlsson, A. H., Hansen-Moller, J., Jensen, M. T., and Tuomola, M. (2006). Influence of chicory roots (Cichorium intybus L) on boar taint in entire male and female pigs. *Animal Science, 82, 359-368.*

[113] Brouns, F., Edwards, S. A., and English, P. R. (1994). Effect of Dietary Fiber and Feeding System on Activity and Oral Behavior of Group-Housed Gilts. *Applied Animal Behavior Science, 39(3-4), 215-223.*

[114] de Leeuw, J. A., Zonderland, J. J., Altena, H., Spoolder, H. A. M., Jongbloed, A. W., and Verstegen, M. W. A. (2005). Effects of levels and sources of dietary fermentable non-starch polysaccharides on blood glucose stability and behaviour of group-housed pregnant gilts. *Applied Animal Behaviour Science, 94(1-2), 15-29.*

[115] Renaudeau, D., Weisbecker, J. L. and Noblet, J. (2003). Effect of season and dietary fibre on feeding behaviour of lactating sows in a tropical climate. *Animal Science, 77, 429-437.*

[116] Gentilini, F. P., Dallanora, D., Bernardi, M. L., Wentz, I., and Bortolozzo, F. P. (2003). Behaviour of pregnant gilts fed low or high fiber diets and maintained in crates or stalls. *Arquivo Brasileiro De Medicina Veterinaria E Zootecnia, 55(5),599-605.*

[117] van der Peet-Schwering, C. M. C., Spoolder, H. A. M., Kemp, B., Binnendijk, G. P., den Hartog, L. A., and Verstegen, M. W. A. (2003). Development of stereotypic behaviour in sows fed a starch diet or a non-starch polysaccharide diet during gestation and lactation over two parities. *Applied Animal Behaviour Science, 83(2), 81-97.*

[118] Meunier-Salaun, M. C., Edwards, S. A., and Robert, S. (2001). Effect of dietary fibre on the behaviour and health of the restricted fed sow. *Animal Feed Science and Technology, 90(1-2), 53-69.*

[119] Danielsen, V. and Vestergaard, E. M. (2001). Dietary fibre for pregnant sows: effect on performance and behaviour. *Animal Feed Science and Technology, 90(1-2), 71-80.*

[120] Bergeron, R., Bolduc, J., Ramonet, Y., Meunier-Salaun, M. C., and Robert, S. (2000). Feeding motivation and stereotypies in pregnant sows fed increasing levels of fibre and/or food. *Applied Animal Behaviour Science, 70(1), 27-40.*

[121] Guilleme, R., Hamard, A., Quesnel, H., Pere, M. C., Etienne, M., Dourmad, J. Y., and Meunier-Salaun, M. C. (2007). Dietary fibre for gestating sows: effects on parturition progress, behaviour, litter and sow performance. *Animal, 1(6), 872-880.*

[122] Shah, P., Nankova, B. B., Patab, S., and La Gamma, E. F. (2006). Short chain fatty acids induce TH gene expression via ERK-dependent phosphorylation of CREB protein. *Brain Research, 1107:13-23.*

[123] Tamminga, S. (1992). Gaseous pollutants produced by farm animal enterprises. In C. Phillips and D. Piggins (Eds.), *Farm animals and the Environment (345-357)*: C. A. B. International.

[124] Aarnink, A. J. A. (1997). Ammonia emission from houses for growing pigs as affected by pen design, indoor climate and behaviour. PhD thesis Animal nutrition group, Wageningen University, Wageningen, The Netherlands.

[125] Dentener, F. J. and Crutzen, P. J. (1994). A 3-Dimensional Model of the Global Ammonia Cycle. *Journal of Atmospheric Chemistry, 19(4), 331-369.*

[126] Freney, J. R., Simson, J. R., and Denmead, O. T. (1983). Volatilisation of ammonia. *Developments in Plant and Soil Science, 9, 1-31.*

[127] Roelofs, J. G. M. and Houdijk, A. L. M. (1991). Ecological effect of ammonia. In V. C. Neilsen, J. H. Voorburg, and P. L'Hermite (Eds),*Odour and ammonia emission from livestock farming (10-16)*, London and New York : Elsevier Applied Science.

[128] Fangmeier, A., Hadwigerfangmeier, A., Vandereerden, L., and Jager, H. J. (1994). Effects of Atmospheric Ammonia on Vegetation - a Review. *Environmental Pollution, 86(1), 43-82.*

[129] Canh, T. T. (1998). Ammonia emission from pig excreta as affected by dietary composition. PhD thesis Animal nutrition group, Wageningen University, Wageningen, The Netherlands.

[130] Canh, T. T., Schrama, J. W., Aarnink, A. J. A., Verstegen, M. W. A., van't Klooster, C. E., and Heetkamp, M. J. W. (1998). Effect of dietary fermentable fibre from pressed sugar-beet pulp silage on ammonia emission from slurry of growing-finishing pigs. *Animal Science, 67, 583-590.*

[131] Canh, T. T., Aarnink, A. J. A., Mroz, Z., Jongbloed, A. W., Schrama, J. W., and Verstegen, M. W. A. (1998). Influence of electrolyte balance and acidifying calcium salts in the diet of growing-finishing pigs on urinary pH, slurry pH and ammonia volatilisation from slurry. *Livestock Production Science, 56(1), 1-13.*

[132] Canh, T. T., Sutton, A. L., Aarnink, A. J. A., Verstegen, M. W. A., Schrama, J. W., and Bakker, G. C. M. (1998). Dietary carbohydrates alter the fecal composition and pH and the ammonia emission from slurry of growing pigs. *Journal of Animal Science, 76(7), 1887-1895.*

[133] Canh, T. T., Aarnink, A. J. A., Verstegen, M. W. A., and Schrama, J. W. (1998). Influence of dietary factors on the pH and ammonia emission of slurry from growing-finishing pigs. *Journal of Animal Science, 76(4), 1123-1130.*

[134] Mroz, Z., Moeser, A. J., Vreman, M., van Diepen, J. T. M., van Kempen, T., Canh, T. T. and Jongbloed, A. W. (2000). Effects of dietary carbohydrates and buffering capacity on nutrient digestibility and manure characteristics in finishing pigs. *Journal of Animal Science, 78(12), 3096-3106.*

[135] Zervas, S. and Zijlstra, R. T. (2002). Effects of dietary protein and fermentable fiber on nitrogen excretion patterns and plasma urea in grower pigs. *Journal of Animal Science, 80(12), 3247-3256.*

[136] Fuller, M. F. and Reeds, P. J. (1998). Nitrogen cycling in the gut. *Annual Reviews Nutrition, 18, 385-411.*

[137] Mosenthin, R., Sauer, W. C., and Ahrens, F. (1994). Dietary Pectins Effect on Ileal and Fecal Amino-Acid Digestibility and Exocrine Pancreatic Secretions in Growing Pigs. *Journal of Nutrition, 124(8),1222-1229.*

[138] Sauer, W. C., Mosenthin, R., Ahrens, F., and Denhartog, L. A. (1991). The Effect of Source of Fiber on Ileal and Fecal Amino-Acid Digestibility and Bacterial Nitrogen-Excretion in Growing Pigs. *Journal of Animal Science, 69(10), 4070-4077.*

[139] Mroz, Z., Jongbloed, A. W., Beers, S., Kemme, P. A., Dejonge, L., Vanberkum, A. K., and Vanderlee, R. A. (1993). Preliminary Studies on Excretory Patterns of Nitrogen and Anaerobic Deterioration of Fecal Protein from Pigs Fed Various Carbohydrates. In M. W. A. Verstegen, L. A. Den hartog, G. J. M. Vankempen, and J. H. M. Metz (Eds.), *Nitrogen Flow in Pig Production and Environmental Consequences (Vol. 69, 247-252)*

[140] Schaefer, J. (1997). Sampling, Characterization and Analysis of Malodours. *Agriculture and Environment, 3(2-3), 121-127.*

[141] Kaufmann, W. (1986). Fermentation in the Forestomachs and the Hind Gut, a Comparison. *Archiv Fur Tierernahrung-Archives of Animal Nutrition, 36(2-3), 205-212.*

[142] Bunce, T. J., Kerley, M. S., Allee, G. L., and Day, B. N. (1995). Feeding fructooligosaccharide to weaned pig improves nitrogen metabolism and reduces odor metabolite excretion. *Journal of Animal Science, 73 (S85), 70.*

[143] Sutton, A. L., Kephart, K. B., Verstegen, M. W. A., Canh, T. T., and Hobbs, P. J. (1999). Potential for reduction of odorous compounds in swine manure through diet modification. *Journal of Animal Science, 77(2), 430-439.*

[144] Gaskins, H. R., Collier, C. T. and Anderson, D. B. (2002). Antibiotics as growth promotants: Mode of action. *Animal Biotechnology, 29-42.*

[145] Thomke, S. and Elwinger, K. (1998). Growth promotants in feeding pigs and poultry. II. Mode of action of antibiotic growth promotants. *Animal Research, 153-167.*

[146] Anderson, D. B., McCracken, V. J., Aminov, R. I., Simpson, J. M., Mackie, R. I., Verstegen, M. W. A., and Gaskins, H. R. (1999). Gut microbiology and growth-promoting antibiotics in swine. *Livestock Feeds and Feeding, 101-108.*

[147] Hathaway, M. R., Dayton, W. R., White, M. E., Henderson, T. L., and Henningson, T. B. (1996). Serum Insulin-Like Growth Factor I (IGF-I) Concentrations Are Increased in Pigs Fed Antimicrobials. *Journal of Animal Science, 1541-1547.*

[148] Hathaway, M. R., Dayton, W. R., White, M. E., Henderson, T. L., Young, D. A., and Doan, T. N. (1999). Effect of feed intake on antimicrobially induced increases in porcine serum insulin-like growth factor I. *Journal of Animal Science, 3208-3214.*

[149] Hariharan, H., Coles, M., Poole, D., and Page, R. (2004). Antibiotic resistance among enterotoxigenic Escherichia coli from piglets and calves with diarrhoea. *Canadian Veterinary Journal, 605-606.*

[150] Hart, W. S., Barton, M. D., and Heuzenroeder, M. W. (2004). Antimicrobial resistance in Campylobacter spp., Escherichia coli and enterococci associated with pigs in Australia. *Journal of Veterinary Medicine Series B: Infectious Diseases and Veterinary Public Health, 216-221.*

[151] O'Hare, C., Doran, G., Delappe, N., Morris, D., Buckley, V., Corbett-Feeney, G., McKeown, P., Anderson, W., and Cormican, M. (2004). Antimicrobial resistance and phage types of human and non-human Salmonella enterica isolates in Ireland, 1998-2003. *Communicable disease and public health / PHLS, 193-199.*

[152] Pena, A., Baeta, L., Silveira, I., Calderon, V., Serrano, C., Reu, C., Sousa, J. C., Peixe, L. (2004). Antibiotic residues in edible tissues and antibiotic resistance of faecal Escherichia coli in pigs from Portugal. *Food Additives and Contaminants, 749-755.*

[153] Aarestrup, F. M., Hasman, H., Jensen, L. B., Moreno, M., Herrero, I. A., Dominguez, L., Finn, M., and Franklin, A. (2002). Antimicrobial resistance among enterococci from pigs in three European countries. *Applied and Environmental Microbiology, 4127-4129.*

[154] Aarestrup, F. M., Rasmussen, S. R., Jensen, N. E., and Artursson, K. (1998). Trends in the resistance to antimicrobial agents of Streptococcus suis isolates from Denmark and Sweden. *Veterinary Microbiology, 71-80.*

[155] Bywater, R., Deluyker, H., Deroover, E., de Jong, A., Rowan, T., Shryock, T., Shuster, D., Thomas, V., Valle, M., Walters, J., Marion, H. and McConville, M. (2004). A

European survey of antimicrobial susceptibility among zoonotic and commensal bacteria isolated from food-producing animals. *Journal of Antimicrobial Chemotherapy. 744-754.*

Reviewed by

Prof. Martin Verstegen,
Animal Nutrition Group, Wageningen University, Wageningen, The Netherlands.
Dr. Jean-Paul Lallès,
Director of Research at INRA Rennes, France

Ligands, Polymers and Amino Acids
Editor: Tsisana Shartava
ISBN: 978-1-61122-793-2

Altering Plant Secondary Metabolism to Achieve Broad Spectrum Insect Control and Reduce Mycotoxins

Eric T. Johnson, Patrick F. Dowd and T. Scott Pinkerton
Crop Bioprotection Research Unit
USDA-Agricultural Research Service
1815 N. University St.
Peoria, IL 61604

Abstract

Mycotoxins are toxic or carcinogenic compounds produced by fungi, including *Aspergilllus* and *Fusarium* molds that colonize seeds of maize, cotton, peanuts or tree nuts. Insect herbivory of plant tissue often enhances mold infection. Transgenic maize expressing the *Bacillus thuringiensis* (Bt) toxin produce ears with lower mycotoxin levels when the target insect pest is effectively controlled. However, Bt toxins are generally species specific and therefore new transgenic strategies for broad insect control need to be developed. The vast diversity of plant secondary biochemicals reflects the strategy of non-motile plants to defend themselves from pathogens and animal herbivores. Research in recent years revealed that many of these plant biochemicals are synthesized by a number of enzymes that are coordinately regulated at the gene level by transcription factors. Altering the expression of these transcription factors by genetic engineering opens the possibility that these secondary biochemicals may be synthesized in specific tissues to combat insect herbivory and reduce mycotoxin contamination. Alternatively, well studied secondary biochemical pathways may be modified or transferred from one plant species to another and tested for effective insect resistance. Transgenic manipulation of secondary metabolism must be carefully considered because redirection of plant energy reserves may hinder plant yield. The advent of genetic engineering and advances in plant biochemistry has opened up the exciting possibilities of utilizing the diversity of secondary biochemicals for protecting valuable crop commodities.

Introduction

Plants synthesize primary metabolites (carbohydrates, lipids and amino acids) which are crucial for their day-to-day survival and serve as chemical building blocks for higher organisms. Plants also produce low molecular weight chemicals, the secondary metabolites, which are also extremely important for the plant's continued existence. Plants have developed chemical defense measures, which generally include secondary metabolites, against pathogens, competitors or herbiores. Nearly 100,000 secondary metabolites have been discovered from the plant kingdom but only half of the structures are known [91]. Based on their biosynthetic pathway, secondary metabolites can be classified into three major groups: the alkaloids, the terpenoids, and the phenylpropanoids and allied phenolic compounds [18]. Terpenoids are derived from the five-carbon precursor isopentenyl diphosphate while the alkaloids, which contain one or more nitrogen atoms, are primarily synthesized from amino acids [18]. The phenolic compounds come from either the shikimic acid pathway or the malonate-acetate pathway [18].

Biosynthesis of secondary metabolites from primary metabolites (such as amino acids) usually requires a number of enzymatic steps. Each of the biosynthetic enzymes is expressed in the cell to complete synthesis of the metabolite. In recent years, studies have documented the presence of metabolic channels in cells where each biosynthetic enzyme in a particular pathway is in close proximity to each other and biosynthetic intermediates can be passed between from one enzyme to the other [56]. Coordinated biosynthesis of secondary metabolite requires a constant level of each enzyme, which is regulated in part at the gene level. Expressing a master regulator (i.e. inducer) of all the genes coding for enzyme of a biosynthetic pathway would ensure that sufficient enzyme is present for secondary metabolite production, and this regulator has been identified for a number of plant secondary metabolite pathways [63;81;118]. However, it is also possible that the biosynthetic pathway is regulated by a rate-limiting enzyme which itself is regulated at the gene or protein (e.g. phosphorylation) level. Genetic engineering of a secondary metabolite is clearly more straightforward with a known master regulator that controls the entire pathway, but modulation of a key enzyme(s) that influences the levels or types of secondary metabolites has been accomplished [22;77;103]. The greatest hindrance to the engineering of secondary metabolites in plants is the limited knowledge of their biosynthetic pathway and incomplete genetic characterization. Others have discussed the recent advances in secondary metabolite engineering [39;118;125] but this review will focus on describing the few studies that have engineered secondary metabolites for effective insect resistance. In addition, the potential for engineering pathways for well characterized secondary metabolites for insect resistance will be discussed.

ANTHOCYANINS AND FLAVONOIDS

While it is generally agreed that anthocyanins are synthesized in floral tissues to attract pollinators [63], there is data to suggest that these pigments have other roles. Some have postulated that the anthocyanins may protect plant tissues from light stress, or serve as antioxidants or function as osmoregulators [16;41;113]. However, others have argued that anthocyanins may be part of a defense strategy against potential herbivores [78;104]. Studies of 2 tree species found that leaf anthocyanins possibly play a role in herbivore protection [58;105]. It may be possible that anthocyanins inhibit larval growth. Some anthocyanins from cotton or petunia flowers can inhibit the development of insect larvae [46;52;53]. Transforming anthocyanin transcription factors into a variety of plants can potentially determine whether they do serve as anti-herbivore molecules. Although the *pap1-D* mutant overexpresses other molecules besides anthocyanins, the growth of first instar fall amyworms was inhibited while eating *pap1-D* leaves compared to wild type leaves [55].

The anthocyanin pathway has been well characterized both at the enzyme and gene level [63]. The studies have demonstrated that anthocyanin and flavonoid biosynthesis is controlled by a small number of master regulators that control multiple genes. For example, in maize, the combination of a basic helix-loop-helix (BHLH) protein (R) and MYB (DNA-binding protein encoded by avian myeloblastosis virus) like protein (C1) control transcriptional activation of anthocyanin biosynthesis [38;76]. This combination of BHLH and MYB proteins for control is not limited to a single plant genus, as the maize R and C1 combination also functions in other plants to induce anthocyanin production [73]. In *Arabidopsis thalliana* (Arabidopsis), a single MYB-like protein (PAP1) induces the production of anthocyanins, lignins and flavonoids in all tissues [12].

The maize MYB protein P1 induces a secondary metabolite pathway distinct from anthocyanins without the need for a BHLH protein [42]. The maize *p1* transcription factor regulates production of 3-deoxy anthocyanins, phlobaphenes (polymers of flavan-4-ols), and C-glycosyl flavones [42]. A number of the biosynthetic steps regulated by *p1* are also utilized by anthocyanin biosynthesis [42]. In silks, *p1* induces the biosynthesis of maysin (Figure 1), a C-glycosyl flavone that inhibits the growth of corn earworm larvae [14]. Corn lines that have high levels of maysin in silks usually have red kernels colored by phlobaphenes, which are generally not preferred by consumers. Transgenic expression of *p1* in silks by a putative silk-specific promoter induced maysin synthesis and increased resistance to first instar corn earworm larvae [54]. However, browning of the transgenic kernels was observed, which indicated that the gene promoter utilized was not tissue specific [54]. This study demonstrated that engineering a transcription factor to produce secondary metabolites (maysin) can increase insect resistance, but tissue specific expression is dependent on effective selective gene promoters.

Isoflavonoids are a group of secondary metabolites produced primarily in legumes, where they are responsible for interactions with microbes [5;101]. Studies have demonstrated that some isoflavonoids act as deterrents, toxins or growth inhibitors to a variety of insects [65;95;107;126;131]. Some isoflavonoids are synthesized by isoflavone synthase from naringenin, which is also a precursor molecule of anthocyanins [72]. The cloning of isoflavone synthase has enabled the metabolic engineering of isoflavones in both legume and other plant species [20;72;111;117]. Induction of the phenylpropanoid pathway in soybean by maize R and C1 transcription factors and concurrent suppression of flavonone 3-hydroxylase (F3H) activity (by expressing the F3H gene in between synthetic inverted repeat sequences) effectively blocked anthocyanin production and increased isoflavone levels [132]. The success of these transgenic plant studies indicate that experiments can be performed to test the effects of elevated levels of isoflavonoids on insect larvae. The non-legume transgenic plants (tobacco, petunia and lettuce) engineered with soybean isoflavone synthase all produced genistein. Genistin (Figure 1), a glucoside of genistein, was implicated in resistance to *Anticarsia gemmatalis*, a pest of soybean [95].

Rutin (quercetin 3-rutinoside, Figure 1), a flavonol glycoside, has deleterious effects on a number of insect species [10;27;49;50;112] but does not appear to affect *Manduca sexta* or *Heliothis virescens* [7]. Experiments have demonstrated that rutin oxidized by peroxidases or tyrosinases (which may occur upon herbivore attack) could be the insect toxic molecule [25]. Chalcone isomerase (CHI) catalyzes the reaction of chalcone to (2S)-naringenin [70]. Levels of rutin in transgenic tomato fruit peel were improved up to 78 fold over control fruit peels by the overexpression of petunia CHI [83]. Overexpression of *Saussurea medusa* (a Chinese medicinal plant) CHI in tobacco plants increased leaf rutin levels [70]. CHI overexpression enhanced floral levels of flavonoids (which includes rutin), but not anthocyanin content, which indicates the presence of an unusual regulatory mechanism [70]. No other flavonols or isoflavones were measured in the CHI overexpressing leaves [70]. Unfortunately, none of the CHI overexpressing tissues mentioned above have been tested for insect resistance.

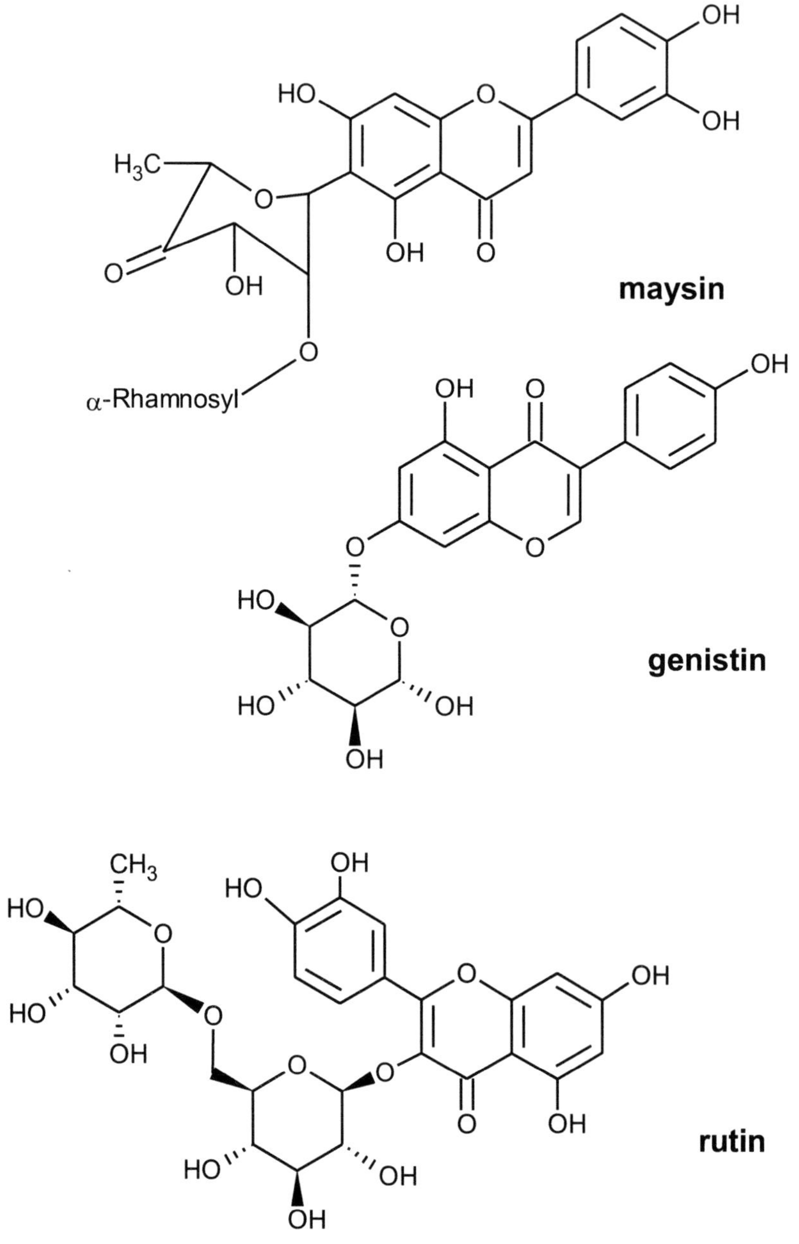

maysin

genistin

rutin

Figure 1. Representative phenolics.

ALKALOIDS

Alkaloids encompass a wide variety of secondary metabolites in 20% of all plants [101]. The current definition includes all nitrogen-containing compounds that are not classified as non-protein amino acids, amines, cyanogenic glycosides, glucosinolates, peptides, cofactors, phytohormones, purines or pyrimidines [101]. Alkaloids are classified into 12 groups based on their chemical structure, and many of them have activity against insects [101]. Unlike phenolic compounds, there is no general or common alkaloid biosynthetic pathway [101]. Thus the entire or partial biosynthetic pathway of an alkaloid with insect bioactivity would need to be engineered into crops depending on whether the crop of interest is capable of synthesizing some types of alkaloids. Biosynthetic pathways of indole (vinblastine, vincristine and captothecin), tropane (atropine, cocaine), pyridine (nicotine), benzylquinoline (morphine, codeine) and benzoxazinoid alkaloids have been fairly well elucidated, but some enzymatic steps are still unknown and very few transcription factors have been identified [39]. In addition, studies determined that biochemical regulation of the indole alkaloid pathway was complex, with intermediates being transported among membrane components, cells and tissues [93]. This makes the prospect of transfer of entire alkaloid pathways to crops that do not produce alkaloids difficult with current knowledge and technology.

The existing understanding of alkaloid biosynthesis and availability of genes can still be exploited for use in crop resistance. Two studies examined the transgenic production of the terpenoid indole alkaloid (TIA) precursor tryptamine (Figure 2) in tobacco and poplar [35;36]. Tryptamine is synthesized from tryptophan by tryptophan decarboxylase [36]. Products of this single gene addition significantly inhibited the growth of two different insect larvae [36]. Though not tested for insect resistance, transgenic petunia plants with elevated tryptamine levels have also been made [116]. These studies demonstrate that elevated levels of some alkaloid biosynthetic intermediates can have detrimental effects on insect growth. Experiments manipulating levels of different alkaloid biosynthetic intermediates for insect resistance may be worth pursuing provided levels of primary metabolites are not significantly compromised.

The benzoxazinoids (Bx) are alkaloids produced in grasses (rye, wheat and maize) that act as defensive compounds against pathogens and insects [86]. 2,4-dihydroxy-7-methoxyl-1,4-benzoxazin-3-one (DIMBOA, Figure 2) is the predominant Bx of maize and is stored as a glucoside in the vacuole [119]. DIMBOA has a broad range of activity, affecting bacteria, fungi and insects such as the corn borer and aphid [80]. When plant tissue is ruptured, DIMBOA glucosides are hydrolyzed by resident β-glucosidases [90] producing DIMBOA aglycones, which have lethal and antifeedant effects on insects [62;87]. A number of studies determined that DIMBOA levels are highest in young seedlings but then decline as the plant ages [8;15;61]. The decline in DIMBOA levels parallels changes in resistance to some maize feeding insects; most strains of maize are resistant to first brood European corn borer larvae as seedlings but most strains lose this resistance as the plant matures [61]. Significant correlations were found between first brood European corn borer larvae resistance and Bx levels in maize [60]. The pathway of DIMBOA biosynthesis has been elucidated and all of the structural genes (*Bx1* through *Bx6*) of the pathway have been cloned [30;31].

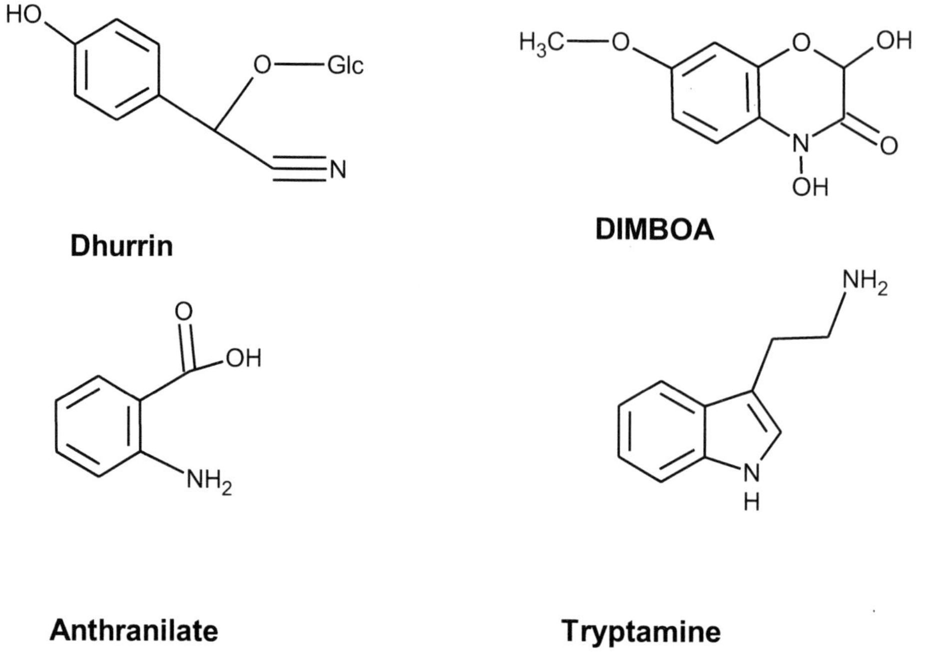

Figure 2. Representative alkaloids, alkaloid precursor, and a cyanogenic glucoside.

It would be technically challenging to transform the entire pathway into a crop using robust promoters for each biosynthetic gene. A more promising approach may utilize genomics to identify possible transcription factors that regulate the entire pathway.

Tryptophan plays an important role in plants not only as an amino acid for protein production, but also as a precursor for indole and quinoline alkaloids [101]. Tryptophan biosynthesis plays a key role in the synthesis of DIMBOA [80]. A key point of metabolic control of tryptophan and related compounds is the enzyme anthranilate synthase (EC 4.1.3.27), which converts chorismate to anthranilate [134]. Anthranilate (Figure 2) is a precursor for acridone and furoquinoline alkaloids that are found mainly in the Rutaceae family [11;99]. Anthranilate synthase enzymatic activity is allosterically controlled by tryptophan levels. Two genes have been identified in *Arabidopsis* as encoding anthranilate synthase and one of those genes, *ASA1*, has been shown to be induced by wounding and bacterial pathogen infiltration [88]. Mutations in anthranilate synthase that can abolish allosteric control can lead to increased amount of tryptophan as well as the accumulation of derivatives of anthranilate. Several plant antranilate synthase mutants have been identified [89;128;129]. Among the phenotypes associated with anthranilate synthase mutations is blue fluorescence due to the accumulation of anthranilate and related compounds. Maize seeds of an L289xI205 cross were exposed to radiation from a nuclear test in 1946 and several mutants were described by Teas and Anderson, including two with a blue fluorescence phenotype [115]. One of these mutants, *bf-1*, is characterized by blue fluorescence in seedlings and anthers under UV illumination, and by volatiles producing a characteristic grape-like odor due to the accumulation of anthranilate derived compounds [84]. The mutant is also resistant to feedback regulation of tryptophan biosynthesis by tryptophan [108]. Sequencing of cDNA from *bf-1* plants showed a point mutation in a conserved chorismate binding region of the anthranilate synthase α subunit which caused a leucine to proline shift at amino acid residue

531 [94]. This mutation is unique compared to other observed mutations in plant anthranilate synthases that convey resistance to feedback inhibition of tryptophan biosynthesis. The *bf-1* anthranilate synthase with an added c-terminal *myc* epitope tag was cloned into the plant expression vector pAHC25 and transferred into maize line *HiII* by particle bombardment [94]. The transgenic plants expressing the *bf-1* mutant anthranilate synthase did not show the strong odor phenotype associated with the *bf-1* mutant plants and only mild fluorescence was observed in some transgenic plants [94]. Callus derived from a transgenic line did show resistance to the anthranilate synthase inhibitory compound 6-methyl anthranilate [94]. Similar blue fluorescent mutants in *Arabidopsis thaliana* have been shown to be dependant on an UDPglucose:salicylic acid glucosyltransferase [97]. Currently we are working to characterize the maize version of this enzyme and another potential modification enzyme, salicylic acid carboxymethyl-transferase, which transfers a methyl group from s-adenosylmethionine to salicylic acid and related compounds (benzoic acid and anthranilic acid) to form methyl esters [67]. The action of these enzymes could explain the presence of both methyl anthranilate and anthranilate-glucose conjugate in the *bf-1* mutant. Work on these enzymes is also of interest because of the recent discovery that methyl salicylate and the salicylic acid carboxymethyltransferase are critical components of the plant systemic acquired resistance pathway in tobacco [92].

Field assays with three different lines of *bf-1* indicated reduced feeding by flea beetles compared to other inbreds and commercial hybrids of the same age growing in the same field (Dowd, unpublished results). Laboratory assays with leaf sections from 3 leaf plants indicated the *bf-1* line tested had significantly less feeding by fall armyworms compared to other lines including its original source hybrid, plus Oh43, Hi-II, I205 and L289 (Dowd, unpublished results). Assays with transformed plants using the mutant *bf-1* gene indicated feeding was significantly inversely correlated with transgenic protein expression levels for corn earworms, and the same trend was noted with fall armyworms [24]. Diet incorporation assays indicated that diet containing 1% methyl anthranilate was significantly active against corn earworms and fall armyworms, causing 100% mortality in both species, and significant feeding reductions occurred when methyl anthranilate was at 0.1% in diet (Pinkerton and Dowd, unpublished results). Although no significant mortality was noted, anthranilic acid caused significant reductions in growth for both corn earworms and fall armyworms at 1% in diet (Pinkerton and Dowd, unpublished results).

TERPENOIDS

Terpenoids are used for a wide range of commercial applications such as flavoring agents, pharmaceutical, perfumes, insecticides and anti-microbial agents [3]. They are synthesized by the melavonate pathway in the cytosol or the Rohmer (non-melavonate) pathway in plastids. Addition of isopentenyl diphosphate groups to dimethyl allyl diphosphate builds geranyl diphosphate (C10), farnesyl diphosphate (C15) or geranylgeranyl diphosphate (C20). Terpene synthases or cyclases then convert these building blocks to the C10 monoterpenes, C15 sesquiterpenes, or C20 diterpenes. Chloroplasts are the site of monoterpene and diterpene biosynthesis while sesquiterpene biosynthesis occurs in the cytosol [3].

Some diterpenes and sequiterpenes kill insects or inhibit their growth [101], but biosynthetic modifications to monoterpene biosynthesis are the easiest to modify by current technologies (see below). Therefore this review will briefly highlight some of the ways in which monoterpenes help plants fight insect attack. Volatile monoterpenes can act as insect repellents. Pure 1,8-cineole is a repellent of mosquitoes, confused flour beetles, and the American cockroach [101;122]. The bicyclic monoterpene isoborneol repels subterranean termites when applied to soil [9]. A number of monoterpenes (e.g. (±)-linalool (Figure 3) and (±)-camphor (Figure 3)) from *Artemisia vulgaris* successfully repelled mosquitoes [51]. Monoterpenoids can also be feeding deterrents. Limonene (Figure 3) and pulegone (Figure 3) are feeding deterrents to cat fleas (*Ctenocephalides felis*) and fall armyworm (*Spodoptera frugiperda*), respectively [101]. Verbenone suppressed the feeding of the pine weevil [71] while both S (+) and R (-) carvone had antifeedant activity against the pales weevil [102].

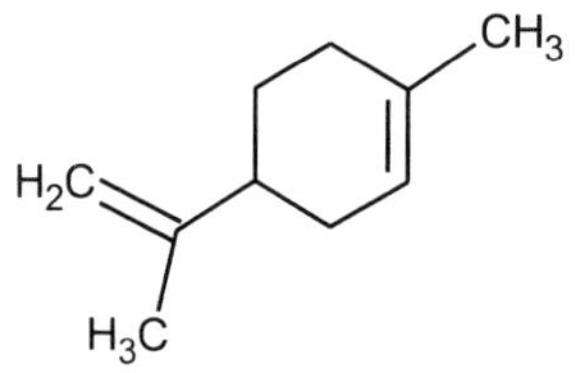

Limonene 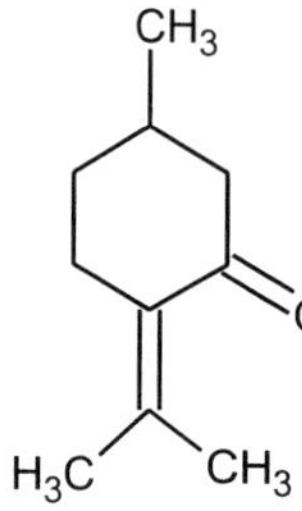 **Pulegone**

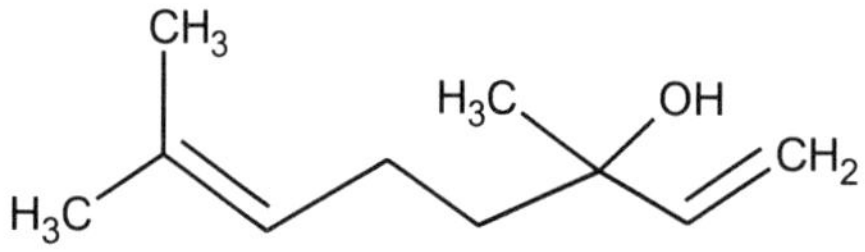

Linalool 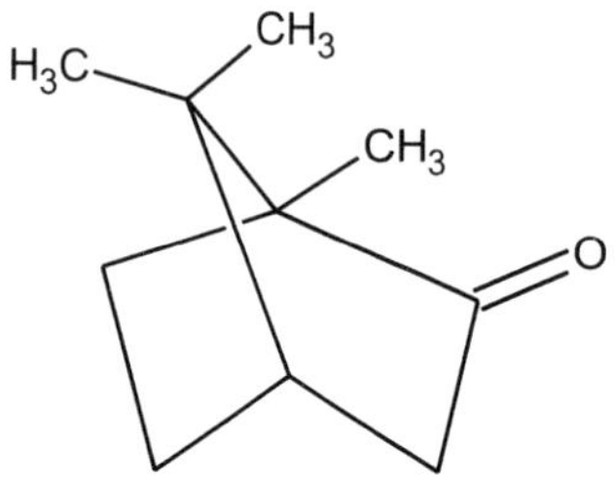 **Camphor**

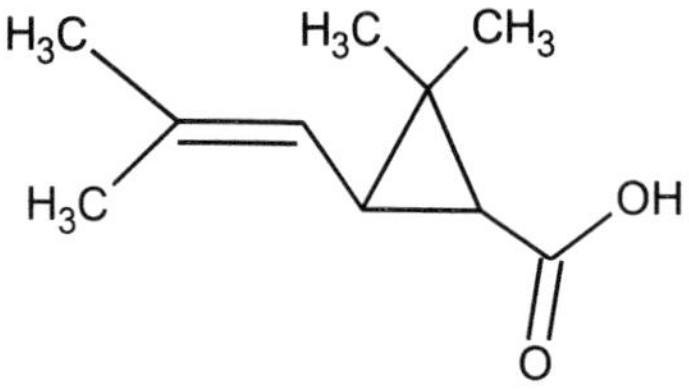

Chrysanthemic acid 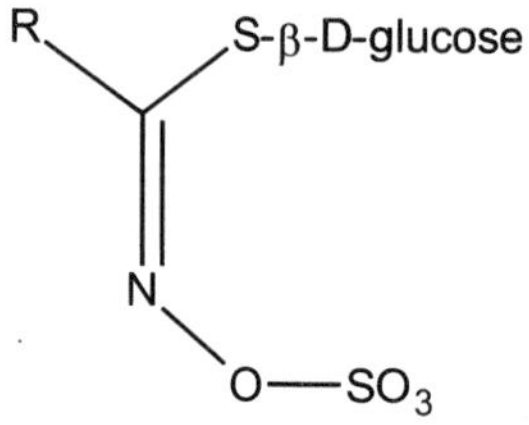 **glucosinolate general structure**

Figure 3. Representative terpenoids and a glucosinolate.

The pyrethroids from *Tanacetum* species are insect neurotoxin terpenoids that affect a wide range of insect species (including caterpillars, beetles and flies) but are rapidly metabolized and excreted in mammals [47].

Pyrethroids are unstable in air and light and thus synthetic pyrethroids (e.g. allethrin) that are more stable have been developed for commercial production [101]. The gene for the first reaction of pyrethroid biosynthesis has been cloned [98], but the genes for the remaining 2 reactions to the insect active compound chrysanthemic acid (Figure 3) have not been isolated. Monoterpenoids, especially pulegone, caused mortality of larvae of *Ostrinia nubialis* when incorporated into artificial diet or when added to the diet surface [68]. Pulegone can also inhibit the larval growth of *Spodoptera eridania* [45].

Efforts to modify terpenoid levels by genetic engineering have been most successful when monoterpene synthases were overexpressed, which increased and/or modified monoterpene profiles [3]. For example, Arabidopsis leaves emit traces of only one monoterpene, limonene [3]. When transformed with a strawberry nerolidol synthase 1 gene targeted to plastids, transgenic Arabidopsis leaves emitted high levels of linalool, in some lines reaching ~7-13 μg day^{-1} plant $^{-1}$ [2]. In addition, the monoterpene profiles of a number of other plants have been modified, including petunia, tobacco, carnation, tomato and potato [3;66;69;74;75]. In the tobacco study, three different lemon monoterpene synthases were transformed into tobacco and then combined into one line by breeding to produce a plant that had higher levels of volatile monoterpenes than wild type plants [75].

Manipulation of monoterpenes for insect resistance is attractive because many important crops, including maize, wheat and rice all synthesize terpenoids and therefore have the basic building blocks of monoterpenoids available in either the cytosol or plastids [17;37;121]. On the other hand, plants that produce high levels of monoterpenes (e.g. mint) store these compounds as oil blends in glandular trichomes [13], presumably to reduce volatility and avoid cytotoxicity. Plants that produce low levels of monoterpenes may have difficulty in storing elevated levels of these compounds. For example, some transgenic potato plants expressing high levels of the strawberry nerolidol synthase 1 gene exhibited bleaching under greenhouse conditions [3]. It remains to be tested whether elevated levels of monoterpenes that preserve plant morphology and yield will have better insect resistance.

A number of volatiles are emitted by plants after attack to attract predators or parasitoids of herbivores, which is termed "indirect defense" [21]. For example, a blend of volatiles, primarily composed of terpenes, is produced by maize after attack by *Spodoptera* larvae, which causes parasitic wasps (*Cotesia marginiventris*) to oviposit on the larvae [123;124]. The blend of maize volatiles varies among maize varieties, which has made identification of the bioactive volatile(s) difficult [40]. Rice plants also emit a blend of sesquiterpenes after treatment with methyl jasmonate, which mimics insect attack [17;79].

The identification of many terpene synthases in recent years has led to the transgenic manipulation of volatile production for indirect defense. The maize terpene synthase clone 10 (*tps10*) catalyzes the conversion of farnesyl diphosphate into a blend of (*E*)-β-farnesene, (*E*)-α-bergamotene, and 7 other minor products in a bacterial system that mirrors the terpene profile of insect damaged maize B73 plants [106]. When *tps10* was transformed into Arabidopsis, the same volatile mixture described above was produced by the rosette leaves and successfully attracted female parasitic wasps (*C. marginiventris*) that had a previous experience of finding host larva on the *tps10* transgenic leaves [106]. Another group transformed Arabidopsis with a strawberry nerolidol synthase gene which produced the homoterpenes (3*S*)-(*E*)-nerolidol and 4,8-dimethyl-1,3(*E*),7-nonatriene in some of the primary transformants [57]. These transgenic plants attracted carnivorous predatory mites (*Phytoseiulus persimilis*) which can benefit plants under herbivore attack [57]. Overexpression of *OsTPS3* (rice sesquiterpene synthase 3) in transgenic rice resulted in more (*E*)-β-caryophyllene after methyl jasmonate treatment which attracted more parasitoid wasps (*Anagrus nilaparvatae*) than wild type plants [17]. While these studies certainly indicate that induced volatile blends can be changed via transgenic methodology to attract beneficial insects, it remains to be determined if indirect defense is a viable strategy in commercial crop protection.

Glucosinolates

These compounds, produced mainly by members of the Brassicaceae, contribute to the unique flavors and aromas of cruciferous vegetables [43]. In addition, glucosinolates are involved in plant defense, auxin homeostasis, and can prevent cancer in humans [43]. The core glucosinolate structure is synthesized from certain amino acids that have a beta-thioglucosyl residue attached to the original alpha-carbon to create a sulfated ketoxime (see Figure 3). Side chain modification and elongation of the amino acid contributes to ~120 known glucosinolate structures [28]. Glucosinolates are normally stored in vacuoles, and cell rupture leads to contact with beta-thioglucosidases called myrosinases [43]. The degradation of glucosinates leads to the creation of bioactive compounds such as isothiocyanates, thiocyanates, nitriles (with β-hydroxylated side chains), and epithionitriles that protect against herbivores and pathogens [101]. For example, isothiocyanates reduce the survival and growth of *Pieris rapae*, an herbivore of plants of the Brassicaceae family, in a dose dependent manner [1]. Glucosinolates also interfere with insect oviposition and egg hatchability, act as phagostimulants for insects adapted to plants of the Brassicaceae, deter feeding of insects, and are toxic to some insects that don't feed on crucifers [101].

Glucosinolates, their biosynthetic pathway, and quantitative trait loci have been extensively studied in Arabidopsis [44;127;135]. Most of the biosynthetic steps of the core glucosinolate structure have been identified [44]. In addition, a number of gene regulators of the pathway have also been identified [32-34;48;109]. If these Arabidopsis regulators function similarly in other species in the Brassicaceae, it may be possible to increase the levels of glucosinolates in important *Brassica* crops like radish, mustard and rapeseed. Transferring the entire core glucosinolate pathway (once it is fully identified) to non-*Brassica* species in the future may be possible. However, synthesized glucosinolates would need to be properly transported to the vacuole for storage and glucosinolate vacuolar transporters have yet not been identified [43]. In addition, a suitable myrosinase gene would also need to be transformed into the non-*Brassica* species and expressed in such a manner that it would not come in contact with glucosinolates. Studies of *Brassica napus* demonstrated that myrosinases were stored in idioblasts called myrosin cells, which were found in all plant organs [4]. Localization of glucosinolates is not as well studied, but it is clear they are stored in vacuoles and not in myrosin cells [4;59]. Strategies for differential storage of glucosinolates and myrosinase need to be identified before glucosinolates can be utilized in plants outside of the Brassicaceae.

CYANOGENIC GLUCOSIDES

Cyanogenic glucosides (CG; see example of dhurrin, Figure 2) are derived from amino acids that release glucose, a ketone or aldehyde, and toxic HCN when hydrolyzed by specific β-glucosidases followed by hydroxynitrile lyase [82;101]. CGs are synthesized in the cytoplasm and stored in the vacuole.

To avoid initiating hydrolysis, the CGs and hydrolyzing enzyme are usually stored in different cells; in sorghum the CGs are stored in epidermal cells while the degrading enzyme is in adjacent mesophyll cells [101]. Hydrolysis can occur under conditions of herbivory, trampling, heat or frost where cellular integrity is compromised [101].

At least 75 different CG have been identified from some 2650 plants, including ferns, gymnosperms, and the angiosperms [101]. Important crop plants containing CGs include sorghum, almond, lima bean, white clover and cassava. Cassava tubers are a major food staple in developing countries. Because the CGs are heat stable, grinding or grating of cassava tubers are necessary for the CGs to come in contact with the β-glucosidase and release the toxic HCN [96]. The free HCN can then be removed by water, cooking, air-drying or sun-drying [96]. CGs are converted to thiocyanates in mammals. Chronic ingestion of CGs coupled with iodine deficiency can lead to goiter and cretinism because thiocyanates interfere with iodine uptake by the thyroid gland [100]. Future biotechnological strategies to utilize CGs for insect resistance in crop plants must consider the necessary post-harvest treatments.

CG adapted insects can sequester CGs, detoxify the released HCN using 3-cyanolalanine synthase or inhibit CG hydrolysis through insect produced β-glycosidases [101]. CGs are toxic to non-adapted herbivores, primarily through cyanide inhibition of cytochrome oxidase and other respiratory enzymes [101]. There is some evidence to suggest that hydrolyzed ketones and aldehydes of CGs can also be toxic [110]. The presence of CGs in different lines of *Trifolim repens* caused various grasshoppers, aphid, slugs, snails and deer to prefer acyanogenic lines [26]. HCN release rates from sorghum leaves depend on plant age, variety differences and other environmental conditions [130]. For example, *Locusta migratoria* will often reject young sorghum leaves but eat older leaves because of differential HCN release rates [130].

Biosynthesis of the CGs have been clearly elucidated in *Sorghum bicolor*, which produces the CG dhurrin (Figure 2) using 3 enzymes that have been genetically identified [6]. The entire dhurrin biosynthetic pathway was engineered into Arabidopsis and the transgenic plants accumulated dhurrin up to 4% dry weight [114]. The cruciferous flea beetle *P. nemorum* was significantly deterred by dhurrin accumulating plants in choice tests comparing wild type tissue [114]. In addition, most flea beetle larvae died while feeding on the dhurrin accumulating Arabidopsis [114]. The three genes for dhurrin biosynthesis were also transformed into grapevine hairy roots that produced up to 100 mg HCN kg^{-1} fresh weight [29]. However, specialist root-sucking insects (*Daktulosphaira vitifoliae*) fed on the cyanogenic transgenic roots and control roots equally well; the authors speculated that higher levels of dhurrin may be necessary for resistance, but only 1 insect species was tested [29]. The successful transformation of the dhurrin biosynthetic pathway from sorghum to two distantly related plants indicates that economically important crops could be transformed as well. The only limitation to development could be the post-harvest treatment necessary for cyanogenic plant tissue that is consumed by humans. Alternatively, the dhurrin biosynthetic pathway could potentially be expressed in tissues (by tissue-specific gene promoters) that are susceptible to insect attack but are consumed in small amounts by humans (e.g. maize husk, silk or tassel).

Secondary Metabolites and Plant Performance

Manipulation of secondary metabolites to enhance insect resistance can potentially hinder other aspects of plant growth and development. However, only a handful of studies have actually measured the metabolic costs of secondary metabolism. As mentioned above, the Arabidopsis *pap1-D* mutant overexpresses a MYB-like transcription factor that results in the induction of anthocyanins, lignins and flavonoids in all tissues [12]. More recent and extensive analysis of around 1800 metabolites of wild type and *pap1-D* mutant plants found that there were only slight differences in the metabolic profile between wild type and the *pap1-D* mutant apart from the substantial increase in anthocyanins and flavonols [120]. The more recent analysis [120] does not indicate changes in lignin accumulation between wild type and *pap1-D* plants while the earlier analysis demonstrated the *pap1-D* mutant produces more lignin [12]. Transcriptome analysis revealed that 38 genes were induced by PAP1 transcription factor [120]. Surprisingly, there were no significant differences in the levels of 16 amino acids, or 12 sugars and anions between wild type and *pap1-D* plants that had been grown in vermiculite for 4 weeks or in sterile culture media for 3 weeks [120]. In addition, there were no changes in levels of phenylalanine between wild type and the *pap1-D* mutant, which indicates that the pools of free phenylalanine were sufficient for the enhanced anthocyanin and flavonol production [120].

Similar results were found for Arabidopsis plants engineered with 3 sorghum genes that synthesized the CG dhurrin up to 4% of the leaf dry weight [64]. The first step of dhurrin biosynthesis utilizes tyrosine, and no significant changes in the free amino acid pools were found between ~30 day old wild type and dhurrin-producing transgenic plants [64]. Analysis of small and global microarrays found very few changes in gene expression between the wild type and dhurrin-transgenic plants [64]. The small metabolic changes observed in the dhurrin transgenic plants may be due in part to the ability of the enzymes of these biosynthetic pathways to form metabolons, which are multiprotein complexes that effectively channel pathway intermediates from one enzyme to the other as discussed above. Recent studies found that the enzymes of the dhurrin biosynthetic pathway do form a metabolon [85]. Arabidopsis plants transformed with only the first 2 genes of dhurrin biosynthesis were stunted and displayed pronounced changes in the transcriptome and metabolome because of the necessary detoxification of *p*-hydroxymandelonitrile, a dhurrin pathway intermediate [64]. In the future, it may be possible to transform crop plants with multistep biosynthetic pathways from heterologous species, as achieved with the dhurrin pathway. Changes to the host transcriptome and metabolome can potentially be minimized by assuring that the donor pathway does not “leak” metabolites and forms a biosynthetic metabolon.

In our studies of the Arabidopsis *pap1-D* mutants, we grew plants with solid purple leaves as well as leaves that only had purple veins, which suggested that some genetic changes had occurred in some progeny of the original mutant [55]. We found the plants with purple-veined leaves produced similar numbers of inflorescences but significantly fewer siliques compared to wild type plants [55]. The solid purple plants produced significantly fewer inflorescences and siliques compared to wild type plants [55]. These findings suggest that the significant production of phenylpropanoid metabolites in the solid purple *pap1-D* plants lowered plant productivity. Perhaps the increased demand for protein deposition in seeds and the continued production of phenylpropanoid metabolites directed by the PAP1 transcription factor begins to deplete the free amino acid pool in late plant development and requires the catabolism of protein or more amino acid synthesis from photosynthate. Based on the number of days (~53) from sowing to flowering of the Arabidopsis Columbia ecotype [19], we presume that the metabolome-analyzed Arabidopsis (Columbia ecotype) wild type and *pap1-D* plants [120], which were metabolically similar apart from anthocyanin and flavonol production, were likely not flowering at the time of leaf analysis. Therefore it remains to be determined if flowering and subsequent seed set significantly alters metabolism in the *pap1-D* mutants. In any case, if secondary metabolism is manipulated in any tissues of crop plants for insect resistance, metabolic studies will need to be measured to ensure that primary metabolism is not compromised.

CONCLUSION

As demonstrated from this short review, there is wide diversity of plant secondary metabolites that can potentially serve as insect resistant molecules. The advantage to studying these secondary metabolites lies in the fact that we are simply trying to identify and utilize the chemical resources that nature has developed over the millennia. The greatest hindrance to utilizing these secondary metabolites is the lack of knowledge of the basic biochemistry of their biosynthesis. Once the biosynthetic enzymes and genes are identified, the pathways may be transformed into model species or the crops themselves. We imagine that as technology improves in the next decade that it will be easier to transform plants with entire biosynthetic pathways. Two metabolic studies of transgenic plants demonstrated that secondary metabolism could be manipulated with very little cost to primary metabolism, but neither of these studies documented the metabolome as the plants shifted their resources into reproduction. Alternatively, if secondary metabolites are identified that are highly toxic to insects and not vertebrates, it may be possible to express them in plants at levels that do not incur a heavy metabolic cost. In addition, some secondary metabolites, such as anthocyanins [133], can benefit human health while potentially reducing insect damage. Reductions in insect herbivory in crops susceptible to fungal colonization (e.g. maize expressing *Bacillus thuringiensis* crystal protein) can result in lower levels of mycotoxins that are harmful to humans [23]. We believe that the utility of secondary metabolite engineering will greatly benefit agriculture in the future.

ACKNOWLEDGMENTS

We thank Mark Berhow and Giltsu Choi for reviewing drafts of the manuscript. We thank Dave Lee for technical assistance.

DISCLAIMER

Mention of trade names or commercial products in this article is solely for the purpose of providing specific information and does not imply recommendation or endorsement by the U.S. Department of Agriculture.

REFERENCES

[1] Agrawal, A. A. and Kurashige, N. S. (2003). A role for isothiocyanates in plant resistance against the specialist herbivore *Pieris rapae*. *J. Chem. Ecol.*, *29*, 1403-1415.

[2] Aharoni, A., Giri, A. P., Deuerline, S., Griepink, F., de-Kogel, W.-J., Verstappen, F. W. A., Verhoeven, H. A., Jongsma, M. A., Schwab, W. and Bouwmeester, H. J. (2003). Terpenoid metabolism in wild-type and transgenic Arabidopsis plants. *Plant Cell*, *15*, 2866-2884.

[3] Aharoni, A., Jongsma, M. A., Kim, T. Y., Ri, M. B., Giri, A. P., Verstappen, F. W. A., Schwab, W. and Bouwmeester, H. J. (2006). Metabolic engineering of terpenoid biosynthesis in plants. *Phytochem. Rev.*, *5*, 49-58.

[4] Andreasson, E., Bolt, J. L., Hoglund, A. S., Rask, L. and Meijer, J. (2001). Different myrosinase and idioblast distribution in Arabidopsis and *Brassica napus*. *Plant Physiol.*, *127*, 1750-1763.

[5] Aoki, T., Akashi, T. and Ayabe, S. (2000). Flavonoids of leguminous plants: structure, biological activity and biosynthesis. *J. Plant Res.*, *113*, 475-488.

[6] Bak, S., Paquette, S., Morant, M., Morant, A. V., Saito, S., Zagrobelny, M., Jørgensen, K., Osmani, S., Jørgensen, B. and Møller, B. L. (2006). Cyanogenic glucosides: A case study for evolution and application of cytochromes P450. *Phytochem. Rev.*, *5*, 309-329.

[7] Bi, J. L., Felton, G. W., Murphy, J. B., Howles, P. A., Dixon, R. A. and Lamb, C. J. (1997). Do plant phenolics confer resistance to specialist and generalist insect herbivores? *J. Agric. Food Chem.*, *45*, 4500-4504.

[8] Bing, J. W., Guthrie, W. D., Dicke, F. F. and Obrycki, J. J. (1990). Relation of corn leaf aphid (Homoptera: Aphididae) colonization to DIMBOA content in maize inbred lines. *J. Econ. Entomol.*, *83*, 1626-1632.

[9] Bläske, V. U., Hertel, H. and Forschler, B. T. (2003). Repellent effects of isoborneol on subterranean termites (Isoptera: Rhinotermitidae) in soils of different composition. *J. Econ. Entomol.*, *96*, 1267-1274.

[10] Bloem, K. A. and Duffey, S. S. (1990). Interactive effect of protein and rutin on larval *Heliothis zea* and the endoparasitoid *Hyposoter exiguae*. *Entomol. Exp. Appl.*, *54*, 149-160.

[11] Bohlmann, J., DeLuca, V., Eilert, U. and Martin, W. (1995). Purification and cDNA cloning of anthranilate synthase from *Ruta graveolens*: modes of expression and properties of native and recombinant enzymes. *Plant J.*, *7*, 491-501.

[12] Borevitz, J. O., Xia, Y., Blount, J., Dixon, R. A. and Lamb, C. (2000). Activation tagging identifies a conserved MYB regulator of phenylpropanoid biosynthesis. *Plant Cell*, *12*, 2383-2394.

[13] Broun, P., Liu, Y., Queen, E., Schwarz, Y., Abenes, M. L. and Leibman, M. (2006). Importance of transcription factors in the regulation of plant secondary metabolism and their relevance to the control of terpenoid accumulation. *Phytochem. Rev.*, *5*, 27-38.

[14] Byrne, P. F., Darrah, L. L., Snook, M. E., Wiseman, B. R., Widstrom, N. W., Moellenback, D. J. and Barry, B. D. (1996). Maize silk browning, maysin content, and antibiosis to the corn earworm, *Helicoverpa zea* (Boddie). *Maydica*, *41*, 13-18.

[15] Cambier, V., Hance, T. and de Hoffman, E. (2000). Variation of DIMBOA and related compounds content in relation to the age and plant organ in maize. *Phytochemistry*, *53*, 223-229.

[16] Chalker-Scott, L. (2002). Do anthocyanins function as osmoregulators in leaf tissues? *Adv. Bot. Res.*, *37*, 103-127.

[17] Cheng, A.-X., Xiang, C.-Y., Li, J.-X., Yang, C.-Q., Hu, W.-L., Wang, L.-J., Lou, Y.-G. and Chen, X.-Y. (2007). The rice (*E*)-β-caryophyllene synthase (OsTPS3) accounts for the major inducible volatile sesquiterpenes. *Phytochemistry*, *68*, 1632-1641.

[18] Croteau, R. B., Kutchan, T. M., and Lewis, N. G. (2000). Natural Products (Secondary Metabolites). In B. B. Buchanan, W. Gruissem, and R. L. Jones (Eds.), *Biochemistry and Molecular Biology of Plants* (pp. 1250-1318). Rockville, MD: American Society of Plant Physiologists.

[19] De Paepe, A., De Grauwe, L., Bertrand, S., Smalle, J. and Van der, S. D. (2005). The *Arabidopsis* mutant *eer2* has enhanced ethylene responses in the light. *J. Exp. Bot.*, *56*, 2409-2420.

[20] Deavours, B. E. and Dixon, R. A. (2005). Metabolic engineering of isoflavonoid biosynthesis in alfalfa. *Plant Physiol.*, *138*, 2245-2259.

[21] Dicke, M. (1999). Evolution of induced indirect defense of plants. In C. D. Harwell, and R. Trollian (Eds.), *The Ecology and Evolution of Inducible Defenses* (pp. 62-88). Princeton, NJ: Princeton University Press.

[22] Diemer, F., Caissard, J. C., Moja, S., Chalchat, J. C. and Jullien, F. (2001). Altered monoterpene composition in transgenic mint following the introduction of 4*S*-limonone synthase. *Plant Physiol. Biochem.*, *39*, 603-614.

[23] Dowd, P. F., Johnson, E. T., and Williams, W. P. (2005). Strategies for insect managment targeted toward mycotoxin management. In H. K. Abbas (Ed.), *Aflatoxin and Food Safety* (pp. 517-541). Boca Raton: CRC Press.

[24] Dowd,P.F., Johnson,E.T., and Pinkerton,T.S. (2007). Selectable markers with potential activity agains insects, plus other insect-oriented strategies for mycotoxin reduction in Midwest corn. *Proceedings of the Multicrop Aflatoxin, Fumonisin and Fungal Genomics Workshop*.

[25] Dowd, P. F. and Vega, F. E. (1996). Enzymatic oxidation products of allelochemicals as a basis for resistance against insects: Effects on the corn leafhopper *Dalbulus maidis*. *Natural Toxins*, *4*, 85-91.

[26] Dritschilo, W., Krummel, J., Nafus, D. and Pimentel, D. (1979). Herbivorous insects colonizing cyanogenic and acyanogenic *Trifolium repens*. *Heredity*, *42*, 49-56.

[27] Elliger, C. A., Wong, Y., Chan, B. G. and Waiss Jr., A. C. (1981). Growth inhibitors in tomato (*Lypersicon*) to tomato fruitworm (*Heliothis zea*). *J. Chem. Ecol.*, *7*, 753-758.

[28] Fahey, J. W., Zalcmann, A. T. and Talalay, P. (2001). The chemical diversity and distribution of glucosinolates and isothiocyanates among plants. *Phytochemistry*, *56*, 5-51.

[29] Franks, T. K., Powell, K. S., Choimes, S., Marsh, E., Iocco, P., Sinclair, B. J., Ford, C. M. and van Heeswijck, R. (2006). Consequences of transferring three sorghum genes for secondary metabolite (cyanogenic glucoside) biosynthesis to grapevine hairy roots. *Transgenic Res.*, *15*, 181-195.

[30] Frey, M., Chomet, P., Glawischnig, E., Stettner, C., Grun, S., Winklmair, A., Eisenreich, W., Bacher, A., Meeley, R. B., Briggs, S. P., Simcox, K. and Gierl, A. (1997). Analysis of a chemical plant defense mechanism in grasses. *Science*, *277*, 696-699.

[31] Frey, M., Huber, K., Park, W. J., Sicker, D., Lindberg, P., Meeley, R. B., Simmons, C. R., Yalpani, N. and Gierl, A. (2003). A 2-oxoglutarate-dependent dioxygenase is integrated in DIMBOA-biosynthesis. *Phytochemistry*, *62*, 371-376.

[32] Gigolashvili, T., Berger, B., Mock, H. P., Muller, C., Weisshaar, B. and Flugge, U. I. (2007). The transcription factor HIG1/MYB51 regulates indolic glucosinolate biosynthesis in *Arabidopsis thaliana*. *Plant J.*, *50*, 886-901.

[33] Gigolashvili, T., Engqvist, M., Yatusevich, R., Muller, C. and Flugge, U. I. (2008). HAG2/MYB76 and HAG3/MYB29 exert a specific and coordinated control on the regulation of aliphatic glucosinolate biosynthesis in *Arabidopsis thaliana*. *New Phytol.*, *177*, 627-642.

[34] Gigolashvili, T., Yatusevich, R., Berger, B., Muller, C. and Flugge, U. I. (2007). The R2R3-MYB transcription factor HAG1/MYB28 is a regulator of methionine-derived glucosinolate biosynthesis in *Arabidopsis thaliana*. *Plant J.*, *51*, 247-261.

[35] Gill, R. I. S. and Ellis, B. E. (2006). Over-expression of tryptophan decarboxylase gene in poplar and its possible role in resistance against *Malacosoma disstria*. *New Forests*, *31*, 195-209.

[36] Gill, R. I. S., Ellis, B. E. and Isman, M. B. (2003). Tryptamine-induced resistance in tryptophan decarboxylase transgenic poplar and tobacco plant against their specific herbivores. *J. Chem. Ecol.*, *29*, 779-793.

[37] Gitelson, I. I., Tikhomirov, A. A., Parshina, O. V., Ushakova, S. A. and Kalacheva, G. S. (2003). Volatile metabolites of higher plant crops as a photosynthesizing life support system component under temperature stress at different light intensities. *Adv. Space Res.*, *31*, 1781-1786.

[38] Goff, S. A., Klein, T. M., Roth, B. A., Fromm, M. E., Cone, K. C., Radicella, J. P. and Chandler, V. L. (1990). Transactivation of anthocyanin biosynthetic genes following transfer of B regulatory genes into maize tissues. *EMBO J.*, *9*, 2517-2522.

[39] Gómez-Galera, S., Pelacho, A. M., Gené, A., Capell, T. and Christou, P. (2007). The genetic manipulation of medicinal and aromatic plants. *Plant Cell Rep.*, *26*, 1689-1715.

[40] Gouinguene, S., Degen, T. and Turlings, T. C. J. (2001). Variability in herbivore-induced odor emissions among maize cultivars and their wild ancestors. *Chemoecology*, *11*, 9-16.

[41] Gould, K. S., McKelvie, J. and Markham, K. R. (2002). Do anthocyanins function as antioxidants in leaves? Imaging of H_2O_2 in red and green leaves after mechanical injury. *Plant, Cell Environ.*, *25*, 1261-1269.

[42] Grotewold, E., Drummond, B. J., Bowen, B. and Peterson, T. (1994). The myb-homologous P gene controls phlobaphene pigmentation in maize floral organs by directly activating a flavonoid biosynthetic gene subset. *Cell*, *76*, 543-553.

[43] Grubb, C. D. and Abel, S. (2006). Glucosinolate metabolism and its control. *Trends Plant Sci.*, *11*, 89-100.

[44] Halkier, B. A. and Gershenzon, J. (2006). Biology and biochemistry of glucosinolates. *Annu. Rev. Plant Biol.*, *57*, 303-333.

[45] Hassanali, A. and Lwande, W. (1989). Antipest secondary metabolites from African plants. *ACS Symp. Series*, *387*, 78-94.

[46] Hedin, P. A., Jenkins, J. N., Collum, D. H., White, W. H., Parrott, W. L. and MacGown, M. W. (1983). Cyanidin-3-β-glucoside, a newly recognized basis for resistance in cotton to the tobacco budworm *Heliothis virescens* (Fab.)(Lepidoptera: Noctuidae). *Experientia*, *39*, 799-801.

[47] Henrick, C. A. (1995). Pyrethroids. In C. R. A. Godfrey (Ed.), *Agrochemicals from Natural Products* (pp. 63-145). New York: Marcel Dekker.

[48] Hirai, M. Y., Sugiyama, K., Sawada, Y., Tohge, T., Obayashi, T., Suzuki, A., Araki, R., Sakurai, N., Suzuki, H., Aoki, K., Goda, H., Nishizawa, O. I., Shibata, D. and Saito, K. (2007). Omics-based identification of *Arabidopsis* Myb transcription factors regulating aliphatic glucosinolate biosynthesis. *Proc. Natl. Acad. Sci. U. S. A.*, *104*, 6478-6483.

[49] Hoffmann-Campo, C. B., Harborne, J. B. and McCaffery, A. R. (2001). Pre-ingestive and post-ingestive effects of soya bean extracts and rutin on *Trichoplusia ni* growth. *Entomol. Exp. Appl.*, *98*, 181-194.

[50] Hoffmann-Campo, C. B., Neto, J. A. R., de Oliveira, M. C. and Oliveira, L. J. (2006). Detrimental effect of rutin on *Anticarsia gemmatalis*. *Pesqui. Agropecu. Bras.*, *41*, 1453-1459.

[51] Hwang, Y.-S., Wu, K.-H., Kumamoto, J., Axelrod, H. and Mulla, M. S. (1985). Isolation and identification of mosquito repellents in *Artemisia vulgaris*. *J. Chem. Ecol.*, *11*, 1297-1306.

[52] Jenkins, J. N., Hedin, P. A., Parrott, W. L., McCarty, J. C. and White, W. H. (1983). Cotton allelochemics and growth of tobacco budworm larvae. *Crop Sci.*, *23*, 1195-1198.

[53] Johnson, E. T., Berhow, M. A. and Dowd, P. F. (2008). Colored and white sectors from star-patterned petunia flowers display differential resistance to corn earworm and cabbage looper larvae. *J. Chem. Ecol.*, in press.

[54] Johnson, E. T., Berhow, M. A. and Dowd, P. F. (2007). Expression of a maize *Myb* transcription factor driven by a putative silk-specific promoter significantly enhances resistance to *Helicoverpa zea* in transgenic maize. *J. Agric. Food Chem.*, *55*, 2998-3003.

[55] Johnson, E. T. and Dowd, P. F. (2004). Differentially enhanced insect resistance, at a cost, in *Arabidopsis thaliana* constitutively expressing a transcription factor of defensive metabolites. *J. Agric. Food Chem.*, *52*, 5135-5138.

[56] Jørgensen, K., Rasmussen, A. V., Morant, M., Nielsen, A. H., Bjarnholt, N., Zagrobelny, M., Bak, S. and Møller, B. L. (2005). Metabolon formation and metabolic channelling in the biosynthesis of plant natural products. *Curr. Opin. Plant Biol.*, *8*, 280-291.

[57] Kappers, I. F., Aharoni, A., van Herpen, T. W., Luckerhoff, L. L., Dicke, M. and Bouwmeester, H. J. (2005). Genetic engineering of terpenoid metabolism attracts bodyguards to *Arabidopsis*. *Science*, *309*, 2070-2072.

[58] Karageorgou, P. and Manetas, Y. (2006). The importance of being red when young: Anthocyanins and the protection of young leaves of *Quercus coccifera* from insect herbivory and excess light. *Tree Physiol.*, *26*, 613-621.

[59] Kelly, P. J., Bones, A. and Rossiter, J. T. (1998). Sub-cellular immunolocalization of the glucosinolate sinigrin in seedlings of *Brassica juncea*. *Planta*, *206*, 370-377.

[60] Klun, J. A., Guthrie, W. D., Hallauer, A. R. and Russell, W. A. (1970). Genetic nature of the concentration of 2,4-dihydroxy-7-methoxy (2*H*)-1,4-benzoxazin-3-(4*H*)-one and resistance to the European corn borer in a diallel set of eleven maize inbreds. *Crop Sci.*, *10*, 87-90.

[61] Klun, J. A. and Robinson, J. F. (1969). Concentration of two 1,4-benzoxazinones in dent corn at various stages of development of the plant and its relation to resistance of the host plant to the European corn borer. *J. Econ. Entomol.*, *62*, 214-220.

[62] Klun, J. A., Tipton, C. L. and Brindley, T. A. (1967). 2,4-dihydroxy-7-methoxy-1,4-benzoxasin-3-one (DIMBOA), an active agent in the resistance of maize to the European corn borer. *J. Econ. Entomol.*, *60*, 1529-1533.

[63] Koes, R., Verweij, W. and Quattrocchio, F. (2005). Flavonoids: a colorful model for the regulation and evolution of biochemical pathways. *Trends Plant Sci.*, *10*, 236-242.

[64] Kristensen, C., Morant, M., Olsen, C. E., Ekstrom, C. T., Galbraith, D. W., Moller, B. L. and Bak, S. (2005). Metabolic engineering of dhurrin in transgenic *Arabidopsis* plants with marginal inadvertent effects on the metabolome and transcriptome. *Proc. Natl. Acad. Sci. U. S. A.*, *102*, 1779-1784.

[65] Lane, G. A., Biggs, D. R., Russell, G. B., Sutherland, O. R. W., Williams, E. M., Maindonald, J. H. and Donnell, D. J. (1985). Isoflavonoid feeding deterrents for *Costelytra zealandica* structure - activity relationships. *J. Chem. Ecol.*, *11*, 1713-1735.

[66] Lavy, M., Zuker, A., Lewinsohn, E., Larkov, O., Ravid, U., Vainstein, A. and Weiss, D. (2002). Linalool and linallol oxide production in transgenic carnation flowers expressing the *Clarkia breweri* linalool synthase gene. *Mol. Breeding*, *9*, 103-111.

[67] Lee, H. I., Leon, J. and Raskin, I. (1995). Biosynthesis and metabolism of salicylic acid. *Proc. Natl. Acad. Sci. U. S. A.*, *92*, 4076-4079.

[68] Lee, S., Tsao, R. and Coats, J. R. (1999). Influence of dietary applied monoterpenoids and derivatives on survival and growth of the European corn borer (Lepidoptera: Pyralidae). *J. Econ. Entomol.*, *92*, 56-67.

[69] Lewinsohn, E., Schalechet, F., Wilkinson, J., Matsui, K., Tadmor, Y., Nam, K. H., Amar, O., Lastochkin, E., Larkov, O., Ravid, U., Hiatt, W., Gepstein, S. and Pichersky, E. (2001). Enhanced levels of the aroma and flavor compound *S*-linalool by metabolic engineering of the terpenoid pathway in tomato fruits. *Plant Physiol.*, *127*, 1256-1265.

[70] Li, F., Jin, Z., Qu, W., Zhao, D. and Ma, F. (2006). Cloning of a cDNA encoding the *Saussurea medusa* chalcone isomerase and its expression in transgenic tobacco. *Plant Physiol. Biochem.*, *44*, 455-461.

[71] Lindgren, B. S., Nordlander, G. and Birgersson, G. (1996). Feeding deterrence of verbenone to the pine weevil, *Hylobius abietis* (L.)(Col., Curculionidae). *J. Appl. Entomol.*, *120*, 397-403.

[72] Liu, R., Hu, Y., Li, J. and Lin, Z. (2007). Production of soybean isoflavone genistein in non-legume plants via genetically modified secondary metabolism pathway. *Metab. Eng.*, *9*, 1-7.

[73] Llyod, A. M., Walbot, V. and Davis, R. W. (1992). *Arabidopsis* and *Nicotiana* anthocyanin production activated by maize regulators *R* and *C1*. *Science*, *258*, 1773-1775.

[74] Lücker, J., Bouwmeester, H. J., Schwab, W., Blaas, J., van der Plas, L. H. and Verhoeven, H. A. (2001). Expression of *Clarkia S*-linalool synthase in transgenic petunia plants results in the accumulation of *S*-linalyl-β-D-glucopyranoside. *Plant J.*, *27*, 315-324.

[75] Lücker, J., Schwab, W., van Hautum, B., Blaas, J., van der Plas, L. H., Bouwmeester, H. J. and Verhoeven, H. A. (2004). Increased and altered fragrance of tobacco plants after metabolic engineering using three monoterpene synthases from lemon. *Plant Physiol.*, *134*, 510-519.

[76] Ludwig, S. R., Bowen, B., Beach, L. and Wessler, S. R. (1990). A regulatory gene as a novel visible marker for maize transformation. *Science*, *247*, 449-450.

[77] Mahmoud, S. S. and Croteau, R. B. (2001). Metabolic engineering of essential oil yield and composition in mint by altering expression of deoxyxylulose phosphate reductoisomerase and menthofuran synthase. *Proc. Natl. Acad. Sci. U. S. A.*, *98*, 8915-8920.

[78] Manetas, Y. (2006). Why some leaves are anthocyanic and why most anthocyanic leaves are red? *Flora: Morph. Dist. Func. Ecol. Plants*, *201*, 163-177.

[79] Martin, D. M., Gershenzon, J. and Bohlmann, J. (2003). Induction of volatile terpene biosynthesis and diurnal emission by methyl jasmonate in foliage of Norway Spruce. *Plant Physiol.*, *132*, 1586-1599.

[80] Melanson, D., Chilton, M. D., Masters-Moore, D. and Chilton, W. S. (1997). A deletion in an indole synthase gene is responsible for the DIMBOA-deficient phenotype of *bxbx* maize. *Proc. Natl. Acad. Sci. U. S. A.*, *94*, 13345-13350.

[81] Memelink, J. and Gantet, P. (2007). Transcription factors involved in terpenoid indole alkaloid biosynthesis in *Catharanthus roseus*. *Phytochem. Rev.*, *6*, 353-362.

[82] Morant, A. V., Jørgensen, K., Jørgensen, B., Dam, W., Olsen, C. E., Møller, B. L. and Bak, S. (2007). Lessons learned from metabolic engineering of cyanogenic glucosides. *Metabolomics*, *3*, 383-398.

[83] Muir, S. R., Collins, G. J., Robinson, S., Hughes, S., Bovy, A., Ric, D., V, van Tunen, A. J. and Verhoeyen, M. E. (2001). Overexpression of petunia chalcone isomerase in tomato results in fruit containing increased levels of flavonols. *Nat. Biotechnol.*, *19*, 470-474.

[84] Neuffer, M. G., Coe, E. H., Jr., and Wessler, S. R. (1997). *Mutants of Maize*. Cold Spring Harbor, NY: Cold Spring Harbor Press.

[85] Nielsen, K. A., Tattersall, D. B., Jones, P. R. and Moller, B. L. (2008). Metabolon formation in dhurrin biosynthesis. *Phytochemistry*, *69*, 88-98.

[86] Niemeyer, H. M. (1988). Hydroxamic acids (4-hydroxy-1,4-benzoxazin-3-ones), defence chemicals in the Gramineae. *Phytochemistry*, *27*, 3349-3358.

[87] Niemeyer, H. M., Pesel, E., Franke, S. and Francke, W. (1989). Ingestion of the benzoxazinone DIMBOA from wheat plants by aphids. *Phytochemistry*, *28*, 2307-2310.

[88] Niyogi, K. K. and Fink, G. R. (1992). Two anthranilate synthase genes in Arabidopsis: defense-related regulation of the tryptophan pathway. *Plant Cell*, *4*, 721-733.

[89] Niyogi, K. K., Last, R. L., Fink, G. R. and Keith, B. (1993). Suppressors of *trp1* fluorescence identify a new Arabidopsis gene, *TRP4*, encoding the anthranilate synthase beta subunit. *Plant Cell*, *5*, 1011-1027.

[90] Oikawa, A., Ebisui, K., Sue, M., Ishihara, A. and Iwamura, H. (1999). Purification and characterization of a β-glucosidase specific for 2,4-dihydroxy-7-methoxy-1,4-benzoxazin-3-one (DIMBOA) glucoside in maize. *Z. Naturforsh.*, *54c*, 181-185.

[91] Oksman-Caldentey, K. M. and Inzé, D. (2004). Plant cell factories in the post-genomic era: new ways to produce designer secondary metabolites. *Trends Plant Sci.*, *9*, 433-440.

[92] Park, S. W., Kaimoyo, E., Kumar, D., Mosher, S. and Klessig, D. F. (2007). Methyl salicylate is a critical mobile signal for plant systemic acquired resistance. *Science*, *318*, 113-116.

[93] Pasquali, G., Porto, D. D. and Fett-Netto, A. G. (2006). Metabolic engineering of cell cultures versus whole plant complexity in production of bioactive monoterpene indole alkaloids: recent progress related to old dilemma. *J. Biosci. Bioeng.*, *101*, 287-296.

[94] Pinkerton,T.S. and Dowd,P.F. (2007). Expression of an anthranilate synthase from maize mutant *bf-1* in maize line Hi II. *Final Program of the Annual Meeting of the American Society of Plant Biologists*, Abstract P45021.

[95] Piubelli, G. C., Hoffmann-Campo, C. B., Moscardi, F., Miyakubo, S. H. and de Oliveira, M. C. (2005). Are chemical compounds important for soybean resistance to *Anticarsia gemmatalis*? *J. Chem. Ecol.*, *31*, 1509-1525.

[96] Poulton, J. E. (1989). Toxic compounds in plant foodstuffs: cyanogens. In J. E. Kinsella, and W. G. Soucie (Eds.), *Food Proteins* (pp. 381-401). Champaign, IL: American Oil Chemists' Society.

[97] Quiel, J. A. and Bender, J. (2003). Glucose conjugation of anthranilate by the *Arabidopsis* UGT74F2 glucosyltransferase is required for tryptophan mutant blue fluorescence. *J. Biol Chem.*, *278*, 6275-6281.

[98] Rivera, S. B., Swedlund, B. D., King, G. J., Bell, R. N., Hussey, C. E., Jr., Shattuck-Eidens, D. M., Wrobel, W. M., Peiser, G. D. and Poulter, C. D. (2001). Chrysanthemyl diphosphate synthase: isolation of the gene and characterization of the recombinant non-head-to-tail monoterpene synthase from *Chrysanthemum cinerariaefolium*. *Proc. Natl. Acad. Sci. U. S. A.*, *98*, 4373-4378.

[99] Rohde, B., Hans, J., Martens, S., Baumert, A., Hunziker, P. and Matern, U. (2008). Anthranilate *N*-methyltransferase, a branch-point enzyme of acridone biosynthesis. *Plant J.*, in press.

[100] Rosling, H. (1987). *Cassava Toxicity and Food Security* (p. 16). Uppsula, Sweden: Tryck kontact.

[101] Sadasivam, S., and Thayumanavan, B. (2003). *Molecular Host Plant Resistance to Pests*. New York: Marcel Dekker, Inc.

[102] Salom, S. M., Gray, J. A., Randall Alford, A., Mulesky, M., Fettig, C. J. and Woods, S. A. (1996). Evaluation of natural products as antifeedants for the pales weevil (Coleoptera: Curculionidae) and as fungitoxins for *Leptographium procerum*. *J. Entomol. Sci.*, *31*, 453-465.

[103] Sato, F., Hashimoto, T., Hachiya, A., Tamura, K., Choi, K. B., Morishige, T., Fujimoto, H. and Yamada, Y. (2001). Metabolic engineering of plant alkaloid biosynthesis. *Proc. Natl. Acad. Sci. U. S. A.*, *98*, 367-372.

[104] Schaefer, H. M. and Rolshausen, G. (2006). Plants on red alert: do insects pay attention? *BioEssays*, *28*, 65-71.

[105] Schlindwein, C. C. D., Feet-Neto, A. G. and Dillenburg, L. R. (2006). Chemical and mechanical changes during leaf expansion of four woody species of a dry Restinga woodland. *Pl. Biol.*, *8*, 430-438.

[106] Schnee, C., Kollner, T. G., Held, M., Turlings, T. C., Gershenzon, J. and Degenhardt, J. (2006). The products of a single maize sesquiterpene synthase form a volatile defense signal that attracts natural enemies of maize herbivores. *Proc. Natl. Acad. Sci. U. S. A.*, *103*, 1129-1134.

[107] Simmonds, M. S. and Stevenson, P. C. (2001). Effects of isoflavonoids from *Cicer* on larvae of *Heliocoverpa armigera*. *J. Chem. Ecol.*, *27*, 965-977.

[108] Singh, M. and Widholm, J. M. (1975). Study of a corn (*Zea mays* L.) mutant (*blue fluorescent-1*) which accumulates anthranilic acid and its beta-glucoside. *Biochem. Genet.*, *13*, 357-367.

[109] Skirycz, A., Reichelt, M., Burow, M., Birkemeyer, C., Rolcik, J., Kopka, J., Zanor, M. I., Gershenzon, J., Strnad, M., Szopa, J., Mueller-Roeber, B. and Witt, I. (2006). DOF transcription factor AtDof1.1 (OBP2) is part of a regulatory network controlling glucosinolate biosynthesis in Arabidopsis. *Plant J.*, *47*, 10-24.

[110] Spencer, K. (1988). K. Spencer (Ed.), *Chemical Mediation of Coevolution* (pp. 167-240). Orlando, FL: Academic Press.

[111] Sreevidya, V. S., Srinivasa, R. C., Sullia, S. B., Ladha, J. K. and Reddy, P. M. (2006). Metabolic engineering of rice with soybean isoflavone synthase for promoting nodulation gene expression in rhizobia. *J. Exp. Bot.*, *57*, 1957-1969.

[112] Stamp, N. E. (1994). Interactive effects of rutin and constant versus alternating temperatures on performance of *Manduca sexta* caterpillars. *Entomol. Exp. Appl.*, *72*, 125-133.

[113] Steyn, W. J., Wand, S. J. E., Holcroft, D. M. and Jacobs, G. (2002). Anthocyanins in vegetative tissues: A proposed unified function in photoprotection. *New Phytol.*, *155*, 349-361.

[114] Tattersall, D. B., Bak, S., Jones, P. R., Olsen, C. E., Nielsen, J. K., Hansen, M. L., Hoj, P. B. and Moller, B. L. (2001). Resistance to an herbivore through engineered cyanogenic glucoside synthesis. *Science*, *293*, 1826-1828.

[115] Teas, H. J. and Anderson, E. G. (1950). Blue fluorescent, a new mutant in maize. *Genetics*, *35*, 696-697.

[116] Thomas, J. C., Akroush, A. M. and Adamus, G. (1999). The indole alkaloid tryptamine produced in transgenic *Petunia hybrida*. *Plant Physiol. Biochem.*, *37*, 665-670.

[117] Tian, L. and Dixon, R. A. (2006). Engineering isoflavone metabolism with an artificial bifunctional enzyme. *Planta*, *224*, 496-507.

[118] Tian, L., Pang, Y. and Dixon, R. A. (2008). Biosynthesis and genetic engineering of proanthocyanidins and (iso)flavonoids. *Phytochem. Rev.*, in press.

[119] Tipton, C. L., Klun, J. A., Husted, R. R. and Pierson, M. D. (1967). Cyclic hydroxamic acids and related compounds from maize. *Biochemistry*, *6*, 2866-2870.

[120] Tohge, T., Nishiyama, Y., Hirai, M. Y., Yano, M., Nakajima, J., Awazuhara, M., Inoue, E., Takahashi, H., Goodenowe, D. B., Kitayama, M., Noji, M., Yamazaki, M. and Saito, K. (2005). Functional genomics by integrated analysis of metabolome and

transcriptome of Arabidopsis plants over-expressing an MYB transcription factor. *Plant J.*, *42*, 218-235.

[121] Ton, J., D'Alessandro, M., Jourdie, V., Jakab, G., Karlen, D., Held, M., Mauch-Mani, B. and Turlings, T. C. J. (2007). Priming by airborne signals boosts direct and indirect resistance in maize. *Plant J.*, *49*, 16-26.

[122] Tunç, I. and Erler, F. (2003). Repellency and repellent stability of essential oil constituents against *Tribolium confusum*. *Z. Pflanzenkr. Pflanzenschutz*, *110*, 394-400.

[123] Turlings, T. C. J., Tumlinson, J. H., Heath, R. B., Proveaux, A. T. and Doolittle, R. E. (1991). Isolation and identification of allelochemicals that attract the larval parasitoid, *Cotesia marginiventris* (Cresson), to the microhabitat of one of its hosts. *J. Chem. Ecol.*, *17*, 2235-2251.

[124] Turlings, T. C. J., Tumlinson, J. H. and Lewis, W. J. (1990). Exploitation of herbivore-induced plant odors by host-seeking parasitic wasps. *Science*, *250*, 1251-1253.

[125] Ververdis, F., Trantas, E., Douglas, C., Vollmer, G., Kretzschmar, G. and Panopoulos, N. (2007). Biotechnology of flavonoids and other phenylpropanoid-derived natural products. Part II: Reconstruction of multienzyme pathways in plants and microbes. *Biotech. J.*, *2*, 1235-1249.

[126] Wang, S. F., Ridsdill-Smith, T. J. and Ghisalberti, E. L. (1999). Levels of isoflavonoids as indicators of resistance of subterranean clover trifolates to redlegged earth mite *Halotydeus destructor*. *J. Chem. Ecol.*, *25*, 795-803.

[127] Wentzell, A. M., Rowe, H. C., Hansen, B. G., Ticconi, C., Halkier, B. A. and Kliebenstein, D. J. (2007). Linking metabolic QTLs with network and cis-eQTLs controlling biosynthetic pathways. *PLoS. Genet.*, *3*, 1687-1701.

[128] Widholm, J. M. (1972). Anthranilate synthetase from 5-methyltryptophan-susceptible and -resistant cultured *Daucus carota* cells. *Biochim. Biophys. Acta*, *279*, 48-57.

[129] Widholm, J. M. (1972). Cultured *Nicotiana tabacum* cells with an altered anthranilate synthetase which is less sensitive to feedback inhibition. *Biochim. Biophys. Acta*, *261*, 52-58.

[130] Woodhead, S. and Bernays, E. (1977). Changes in release rates of cyanide in relation to palatability of sorghum to insects. *Nature*, *270*, 235-236.

[131] Youn, H., Lee, S., Cho, J. H. and Oh, H. (1991). Insecticidal isoflavon glycoside from *Maackia amurensis*. *Arch. Pharm. Res.*, *14*, 105-108.

[132] Yu, O., Shi, J., Hession, A. O., Maxwell, C. A., McGonigle, B. and Odell, J. T. (2003). Metabolic engineering to increase isoflavone biosynthesis in soybean seed. *Phytochemistry*, *63*, 753-763.

[133] Zafra-Stone, S., Bagchi, M. and Bagchi, D. (2007). Health benefits of edible berry anthocyanins: novel antioxidant and anti-angiogenic properties. *ACS Symp. Series*, *956*, 337-351.

[134] Zalkin, H. (1985). Anthranilate synthase. *Methods Enzymol.*, *113*, 287-292.

[135] Zhang, Z., Ober, J. A. and Kliebenstein, D. J. (2006). The gene controlling the quantitative trait locus *EPITHIOSPECIFIER MODIFIER1* alters glucosinolate hydrolysis and insect resistance in *Arabidopsis*. *Plant Cell*, *18*, 1524-1536.

INDEX

A

B

C

D

E

F

G

H

I

J

K

L

M

N

O

P

Q

T

U

V

W

X

Y

Z